Rare Earth Elements and Their Minerals

*Edited by Michael Aide
and Takahito Nakajima*

Published in London, United Kingdom

IntechOpen

Supporting open minds since 2005

Rare Earth Elements and Their Minerals
http://dx.doi.org/10.5772/intechopen.77602
Edited by Michael Aide and Takahito Nakajima

Contributors
Rasha El-Mekawy, Takahito Nakajima, Oyunbold Lamid-Ochir, Igor V. Shamanin, Mishik A. Kazaryan, Raunak K Tamrakar, Kanchan Upadhyay, Miloš René, Michael Thomas Aide

Notice
Statements and opinions expressed in the chapters are these of the individual contributors and not necessarily those of the editors or publisher. No responsibility is accepted for the accuracy of information contained in the published chapters. The publisher assumes no responsibility for any damage or injury to persons or property arising out of the use of any materials, instructions, methods or ideas contained in the book.

First published in London, United Kingdom, 2020 by IntechOpen
IntechOpen is the global imprint of INTECHOPEN LIMITED, registered in England and Wales, registration number: 11086078, 7th floor, 10 Lower Thames Street, London,
EC3R 6AF, United Kingdom
Printed in Croatia

British Library Cataloguing-in-Publication Data
A catalogue record for this book is available from the British Library

Additional hard and PDF copies can be obtained from orders@intechopen.com

Rare Earth Elements and Their Minerals
Edited by Michael Aide and Takahito Nakajima
p. cm.
Print ISBN 978-1-78984-740-6
Online ISBN 978-1-78984-741-3
eBook (PDF) ISBN 978-1-83880-990-4

We are IntechOpen,
the world's leading publisher of
Open Access books
Built by scientists, for scientists

4,900+

Open access books available

123,000+

International authors and editors

140M+

Downloads

151

Countries delivered to

Our authors are among the

Top 1%

most cited scientists

12.2%

Contributors from top 500 universities

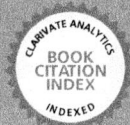

Interested in publishing with us?
Contact book.department@intechopen.com

Numbers displayed above are based on latest data collected.
For more information visit www.intechopen.com

Meet the editors

Dr. Michael Aide received his Ph.D. in soil chemistry from Mississippi State University (1982) and a baccalaureate degree in chemistry and mathematics from the University of Wisconsin – Madison. He has been an educator and agronomic researcher at Southeast Missouri State University since 1982. His research interests involve the growth and development of rice in integrated systems involving soil fertility, water management, and integrated pest management. Rice research has permitted his travel to southeastern Asia, Central America, and Egypt. Dr. Aide has also investigated the use of rare earth elements in soils, particularly attempting to utilize them to indicate parent material uniformity and their fate/transport. Professional affiliations include the American Society of Agronomy, the Soil Science Soc. America, and he is a certified professional soil scientist.

Takahito Nakajima, MD, Ph.D. is a radiologist with a career of over 20 years. He works at Gunma University Hospital and is the Vice Chair of the Radiology Department. His interest is in diagnostic radiology, including CT, MRI, PET, nuclear medicine, and ultrasound sonography, ranging from basic science to clinical studies. In basic research, he has engaged in in-vivo fluorescence imaging, in particular, activatable fluorescence and near-infrared based photoimmunotherapy. He has a long history of working in the clinical field and contributes to quantitative imaging of various organs, with contrast agents being one of his research fields.

Contents

Preface

The book "Rare Earth Elements and Their Minerals" presents a compelling insight into the rapidly evolving discipline of rare earth elements (REE) and their technology utilization. The chapter "Organic and Inorganic Rare Earth Element Hydrolysis and Complexation in Surface and Groundwater: A Review with Chemical Thermodynamic Simulations" observes the complexity of understanding REE in the aqueous environment, focusing on transport potential, biological availability, and system reactivity towards changes in pH, oxidation-reduction, chemical composition, mineralogy (facies) changes, or anthropogenic alteration. Similarly, the chapter "Allanite from Granitic Rocks of the Moldanubian Batholith (Central European Variscan Belt)" demonstrates the complexity of REE accumulation as a mineable ore material and the global search for these valuable deposits.

Subsequent chapters focus on REE synthetic analogues for use in advanced technologies and medical applications. The chapter "An Efficient Route for Synthesis of Macrocyclic Gadolinium Complex and Their Role in Medical Applications" observed the coordination chemistry of gadolinium as applied to contrast agents for usage in magnetic resonance imaging. Similarly, the chapter "Current Clinical Issues: Deposition of Gadolinium Chelates" focuses on the effects of gadolinium-chelate based contrast agents on nephrogenic systemic fibrosis and brain deposition. The chapter "Application of the Gadolinium Isotopes Nuclei Neutron-induced Excitation Process" investigates the creation of laser technology for pulse power engineering. Lastly the chapter "Gd_2O_3: A Luminescent Material" researched the structural and optical behavior of a Gd_2O_3 nanophosphor to detail its photoluminescence to position its usage as a phosphor for dosimetric applications.

Dr. Michael Aide
Professor of Agriculture,
Southeast Missouri State University,
Missouri, USA

Dr. Tahito Nakajima
Gunma University Hospital,
Gunma, Japan

Rare Earth Elements and the Environment

Review and Assessment of Organic and Inorganic Rare Earth Element Complexation in Soil, Surface Water, and Groundwater

Michael Aide

Abstract

The lanthanide elements, or rare earth elements (REEs), are an active research area, with increasing attention directed toward soil and water evaluation and protection. Rare earth element concentrations in surface and groundwaters may be appropriately evaluated by partitioning the REEs into (i) a dissolved fraction (REE^{3+}, hydrolysis, and simple anion complexation products) and (ii) REEs associated with inorganic and organic colloidal fractions. Given the total REE concentration and the organic, inorganic, and clastic composition, each fraction of REE concentration and the speciation within the fraction may be thermodynamically simulated to estimate (i) transport potential, (ii) biological availability, and (iii) system reactivity toward changes in pH, oxidation-reduction, chemical composition, mineralogy (facies) changes, or anthropogenic alteration. Chemical thermodynamic simulations using freely available USEPA software are presented to illustrate REE alterations attributed to pH changes, inorganic and organic adsorption, mineral precipitation, and oxidation-reduction. The purpose is to position researchers to better anticipate REE reactivity and transport potential in aquatic and soil resources.

Keywords: lanthanides, Minteq, aquatic resources, environmental degradation, thermodynamic simulation

1. Rare earth element: inorganic chemistry

The rare earth elements (REEs) are the 14 elements comprising the lanthanide series: cerium (Ce), praseodymium (Pr), neodymium (Nd), promethium (Pm), samarium (Sm), europium (Eu), gadolinium (Gd), terbium (Tb), dysprosium (Dy), holmium (Ho), erbium (Er), thulium (Tm), ytterbium (Yb), and lutetium (Lu) [1]. Lanthanum (La) is associated with the rare earth elements because of its trivalent chemical affinity and periodic table position. The lanthanide series are elements characterized as having one or more electrons in the 4f electronic orbitals for their ground state configuration. Promethium undergoes rapid radioactive decay and is absent in the environment.

The typically trivalent REE elements have considerable ionic bonding character [1]. The chemical attributes of the REEs are influenced by the regular decrease in the ionic radii on progression from La to Lu, the so-called lanthanide contraction. The "lanthanide contraction" occurs because of the incomplete electric field shielding by the f orbitals and increases in atomic number, supporting greater chemical affinity for hydrolysis and chelate/complex stability on progression across the lanthanide series [1]. The LREE are the light rare earth elements, comprised of the elements La to Eu, and the HREE are the heavy rare earth elements, comprised of the elements Gd to Lu. In some cases, The REEs have been partitioned as (i) the light REE (LREE includes La, Ce, Pr), (ii) the middle REE (MREE includes Nd, Sm, Eu, and Gd), and (iii) the heavy REE (HREE includes Tb to Lu).

2. Hydrolysis and complexation thermodynamic data

The hydrolysis of REE^{3+} species has been extensively investigated. The primary thermodynamic literature featuring data involving REE^{3+} hydrolysis and inorganic complexation reactions include Baes and Mesmer [2], Hummel et al. [3], Smith and Martel [4], Schijf and Byrne [5], Luo and Byrne [6], Cantrell and Byrne [7], Gramaccioli et al. [8], Lee and Byrne [9], and Millero [10]. Klungness and Byrne [11] noted that REE hydrolysis is more stable with increasing atomic number across the lanthanide series.

Inorganic complexation of the REE elements involves coordination with primarily anionic species, and it is expressed as.

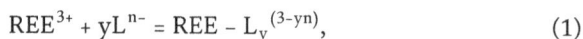

$$REE^{3+} + yL^{n-} = REE - L_y^{(3-yn)}, \tag{1}$$

where L^{n-} is an inorganic ligand with n ionic charge and y is the stoichiometric coefficient. For the lanthanide series, the dicarbonate complex becomes increasingly more stable with increasing atomic number [6, 7, 9]. Both hydrolysis and carbonate complexation show the expected increasing stability with increasing atomic number across the lanthanide series [12]. Aide [12] reviewed thermodynamic data concerning rare earth element hydrolysis.

Common low-molecular-weight organic complexes include acetic acid, phthalic acid, oxalic acid, lactic acid, malic acid, and citric acid. Humus components typically include fulvic and humic acids. The seminal literature featuring thermodynamic data involving REE^{3+} organic complexation include Gu et al. [13], Dong et al. [14], and Pourret et al. [15].

3. Distribution of rare earth elements in soils and earth materials

REE concentrations in soils, sediments, and other earth materials are dependent on their mineral assemblage and source area, with REE concentrations typically ranging from 0.1 to 100 mg/kg. In general, felsics have greater REE concentrations and greater LREE/HREE ratios than mafics. As expected, fine-grained clastic sediments frequently exhibit greater REE concentrations than limestones and sandstones. The Oddo-Harkins rule states that an element with an even atomic number has a greater concentration than the next element in the periodic table. The REEs typically obey the Oddo-Harkins rule. The Post-Archean Australian average shale (PAAS), North American shale composite (NASC), selective representative soil collections, and selected geochemical soil surveys usually reflect the Oddo-Harkins rule [16–18] (**Table 1**).

Commonly occurring REE-bearing minerals include (i) fluorite (Ce replaces Ca), (ii) allanite (Ce), (iii) sphene (REE replace Ca), (iv) zircon (HREE replace Zr), (v) apatite (REE replace Ca), (vi) monazite ((CeLa) phosphate),

Element	PAAS[1]	NASC[1]	Soil[2]	Chinese soils[3]
			mg/kg	
La	38.2	32	26.1	37.4
Ce	79.6	73	48.7	64.7
Pr	8.83	7.9	7.6	6.67
Nd	33.9	33	19.5	25.1
Sm	5.55	5.7	4.8	4.94
Eu	1.08	1.24	1.2	0.98
Gd	4.66	5.2	6.0	4.38
Tb	0.774	0.85	0.7	0.58
Dy	4.68	5.8	3.7	3.93
Ho	0.991	1.04	1.1	0.83
Er	2.85	3.4	1.6	2.42
Tm	0.405	0.5	0.5	0.35
Yb	2.82	3.1	2.1	2.32
Lu	0.433	0.48	0.3	0.35

[1]Reported in McLennan [16].
[2]Reported in Kabata-Pendias [17].
[3]Reported in Liang et al. [18].
PAAS is Post-Archean Australian average shale; NASC is North American shale composite.

Table 1.
Rare earth element abundances for various parent materials.

(vii) xenotime (REE—phosphate), (viii) rhabdophane (Ce, REE—phosphate), and (ix) bastnaesite (REE fluorocarbonate). As with many mineral assemblies, the soil LREE concentrations are generally greater than the soil HREE; however, mineral assemblages featuring an abundance of zircon may differ in the LREE/HREE.

4. Rare earth element abundances in natural waters: river water

Natural waters include marine, river, lacustrine, and groundwater. Considerations for characterizing natural water REE concentrations include (1) the total REE concentration; (2) suspended minerals having adsorbed, occluded, or latticed REE; (3) organically complexed REE; and (4) soluble REE^{3+} and their hydrolytic and ion pair products. Liang et al. [18] cited literature references for river waters in China. The REE distribution shows that the light rare earth elements (La to Eu) are more abundant than the heavy REEs (Gd to Lu) and the distribution follows the Oddo-Harkins rule. These authors also compared rivers having either pristine and REE impaction because of REE mining activities (**Table 2**). The REE concentrations because of mining activity were intense, underscoring the environmental impact. Linear regression by the author of this manuscript of Liang et al. [18] river water data for dissolved and suspended REE load shows substantial correlation between the dissolved and suspended concentrations for all REEs. The linear relationship for the Yellow River segment [suspended matter = 317.86 (dissolved) −4.904 with r^2 = 0.91] and the Kundulum River segment [suspended matter = 0.787 (dissolved) +0.624 with r^2 = 0.95] infer that the respective suspended and dissolved REE concentrations arise from similar chemical adsorption relationships.

	Dissolved (µg L^{-1})		Suspended matter (µg g^{-1})	
	Yellow	Kundulum	Yellow	Kundulum
La	0.10	140	33.16	86.16
Ce	0.22	152	68.26	139.3
Pr	0.034	16.1	8.47	15.8
Nd	0.095	52.0	28.7	52.0
Sm	0.054	6.91	5.79	6.61
Eu	0.018	1.52	1.31	1.20
Gd	0.028	7.22	6.29	4.31
Tb	0.006	0.88	0.82	0.97
Dy	0.08	3.80	3.39	1.87
Ho	0.014	0.18	0.77	0.68
Er	0.03	2.34	2.25	1.36
Tm	0.007	0.24	0.25	0.22
Yb	0.032	1.48	1.76	1.01
Lu	0.006	0.21	0.25	0.22

Documented in Liang et al. [18].

Table 2.
Rare earth element concentrations documented for two Chinese rivers.

River waters typically have greater REE concentrations than marine waters because of their suspended load and a greater abundance of dissolved organic material [19–22]. Dupre et al. [19] observed that the REEs were primarily associated with suspended inorganic and organic colloids. Garcia et al. [20] studied river waters in Argentina draining predominately granitic landscapes, showing that high-rainfall periods effectively reduced or "diluted" REE concentrations. Andersson et al. [21] proposed that organic colloidal materials were largely responsible for REE transport in boreal Swedish river waters and that the LREE were more abundant than the HREE. Ingri et al. [22] demonstrated that the La concentrations in Swedish boreal river waters were seasonal and were associated with organic and Fe-oxyhydroxide inorganic colloidal material.

Gurumurthy et al. [23] documented 3 years of river discharge across southwestern India and provided river water chemistry, including rare earth elements (**Table 3**). They observed that the rare earth elements showed higher concentrations during the monsoon season as opposed to the dry season, suggesting that soil leaching across the watersheds was important to the increased monsoonal river water concentrations. Cerium anomalies were observed, pH moderated adsorption-desorption reactions, and the dissolved oxygen concentrations were important in regulating the seasonality of the Ce anomalies. Rare earth element complexation was not highly significant in influencing the rare earth element concentrations.

Neal [24] documented La, Ce, Pr, and yttrium (Y) concentrations in the upper River Severn catchments in Mid Wales. Over a 7-year interval, larger river water concentrations of La, Ce, Pr, and Y were associated with high-rainfall events and baseflow/return flow from land parcels having acidic soil pH values, suggesting that the surrounding terrestrial environment is important to REE river chemistry. Leybourne and Johannesson [25] described that the REE adsorption affinity for stream waters and sediments was pH-dependent, with deprotonation of surface

Southwestern Indian rivers (pmol L^{-1})				
Element	Bantwal	Gurupur	Mugeru	Shanthimugeru
La	854	793	1213	1136
Nd	759	765	1130	1096
Ce	1986	1867	2698	2658
Pr	198	189	291	276
Sm	162	159	216	229
Eu	46	44	62	62
Gd	148	134	209	211
Dy	112	105	166	174
Ho	27	28	34	36
Er	65	61	90	100
Tm	11	19	14	16
Yb	57	57	74	89

Documented in Gurumurthy et al. [23].

Table 3.
Rare earth elements discharge-weighted mean averages of rare earth elements in river waters from southwestern India.

hydroxyl groups favoring REE adsorption at more alkaline pH intervals. With increasing pH, the adsorption potential may permit REE fractionation, with the adsorption affinity greatest for the LREE, less for the MREE, and least for the HREE. In Sweden, Ohlander et al. [26] recorded Sm/Nd ratios in the weathering of granitic till, noting Sm/Nd differences in the upper eluvial soil horizons relative to the deeper less weathered till. Weathering intensity differences and secondary preferential placement of Nd in the deeper less weathered till influenced adjacent stream compositions.

5. Rare earth element abundances in natural waters: groundwater

The total rare earth element concentrations in groundwater may be partitioned into (i) dissolved or free ion species that may include hydrolysis products and inorganic complexes, (ii) low-molecular-weight organic ligands and moderate- to large-molecular-weight chelates (e.g., humic and fulvic acids (FA)), and (iii) clastic colloids (e.g., phyllosilicates and Fe-oxyhydroxides) [27–33]. Groundwater may frequently exhibit a seasonal range in total REE concentrations [27]. Dia et al. [27] documented REE, dissolved organic carbon (DOC), and trace metals in well waters from a French catchment, noting that spatially distinct groundwaters may be partitioned based on DOC content and other hydrologic variables. Ultrafiltration of the distinct groundwaters reveals that the REE concentrations in the organic-rich waters were more associated with organic colloids, whereas the REEs in groundwaters having small DOC concentrations were more associated with inorganic colloids. Similarly, Pourret et al. [28], working with the same catchment as Dia et al. [27], employed ultrafiltration techniques and species modeling using the humic ion-binding model VI to show that (i) the smaller REE concentrations in ultrafiltration waters were attributed to the removal of REE-bearing organic colloids and (ii) modeling suggests that the lanthanum complexes were dominated by humic acids (80%) and subordinately with fulvic acids (20%). Inorganic complexes were of greater importance in groundwaters having low DOC concentrations. Omonona and Okoghue [31] showed REE concentrations from Nigerian aquifers, demonstrating the region's water REE chemical diversity (**Table 4**).

Adsorption reactions involving the REEs and aquifer materials are instrumental to understanding REE water concentrations and transport [34–41]. Rabung et al. [34] performed batch adsorption experiments involving Eu^{3+} on Ca-montmorillonite and Na-illite and showed Eu outer-sphere complexes at pH levels less than pH 4 on illite, whereas no outer-sphere complexes were observed with montmorillonite. For pH levels greater than pH 5, inner-sphere complexes were formed for both minerals. Coppin et al. [29] showed that lanthanide adsorption on smectite and kaolinite was pH and ionic strength dependent and demonstrated increased adsorption at higher ionic strengths near pH 5.5. At lower ionic strengths, REE adsorption onto smectite was weakly pH-dependent from 3 to pH 6, whereas REE adsorption was increasingly greater above pH 6. Kaolinite showed increased REE adsorption with increased pH. At the greater ionic strength, the heavy REEs exhibited greater adsorption, a feature consistent with lanthanide contraction.

Cteiner [42] observed monazite ($NdPO_4$) reactivity at low ionic strengths to estimate the influence of Cl^-, HCO_3^-, SO_4^{2-}, oxalate, and acetate on monazite solubility. At pH levels ranging from 6.0 to 6.5, Nd (oxalate) was the dominant species, followed by Nd^{3+} and $NdSO_4^+$. Davranche et al. [37, 38] demonstrated that REEs and humic acid complexes frequently dominate soil aqueous systems, especially in near-neutral pH levels and at greater dissolved organic carbon concentrations. Pourret et al. [43] observed the strong competitive interaction between humic acids and carbonates for REE complexation, especially at increasing pH levels. Similarly, Wu et al. [36] described the strong competition involving EDTA and humic and fulvic acids, which effectively inhibited lanthanum adsorption onto goethite.

Cation exchange and adsorption reactions involving cations and their hydrolytic products are dominant soil processes, including (i) multi-site cation exchange reactions, (ii) adsorption reactions with increasing degree of inner-sphere complexes

Element	Low	High	Mean
		($\mu g\ L^{-1}$)	
La	0.33	42.85	6.83
Ce	0.73	85.15	6.83
Nd	0.36	36.51	6.18
Pr	0.09	9.25	1.55
Sm	0.05	5.47	1.04
Eu	0.00	0.50	0.07
Gd	0.06	3.61	0.81
Dy	0.00	2.08	0.49
Ho	0.00	0.38	0.09
Er	0.01	0.94	0.23
Tm	0.00	0.12	0.03
Yb	0.00	0.80	0.18

Source: Omonona and Okoghue [31].

Table 4.
Rare earth element concentrations from selected aquifers in the Gboko area, Nigeria.

at pH levels greater than pH 5, (iii) REE affinity being reduced by increased ionic strength, and (iv) REE complexation affinity being greater at higher pH intervals. Davranche et al. [38] provided adsorption data on hydrous ferric oxides with REEs and REE-humate complexes. REE-humate complexes do not dissociate upon adsorption, with binding presumed to be anionic adsorption involving the humate portion of the complex. Pourret et al. [15] employed ultrafiltrate techniques to investigate La, Eu, and Lu synthetic humic acid complexation and modeled the datasets with the humic ion-binding model to demonstrate that the quantity of REE bonding increases with pH. The intensity of the REE-humic acid binding approached 100% near pH 4 for the highest humic acid concentration (20 mg L^{-1}). Rare earth complexes involving carbonate and especially dicarbonate speciation were effective competing anions in alkaline media with the effectiveness of carbonate complexation increasing from La to Lu.

6. Materials and methods

An aqua regia digestion was employed to obtain a near total estimation of elemental abundance associated with all but the most recalcitrant soil chemical environments. Aqua regia does not appreciably degrade quartz, albite, orthoclase, anatase, barite, monazite, sphene, chromite, ilmenite, rutile, and cassiterite; however, anorthite and phyllosilicates are partially digested. Homogenized samples (0.75 g) were equilibrated with 0.01 L of aqua regia (3 mole nitric acid/1 mole hydrochloric acid) in a 35°C incubator for 24 hours. Samples were shaken, centri-fuged, and filtered (0.45 μm), with a known aliquot volume analyzed using induc-tively coupled plasma mass spectrometry (ICP-MS).

A hot water extraction was performed to recover only the most labile or poten-tially labile fractions. A hot water extraction involved equilibrating 0.5 g samples in 0.02 L distilled-deionized water at 80°C for 1 hour followed by 0.45 μm filtering and elemental determination using ICP-MS. In the water extract and the aqua regia digestion, selected samples were duplicated, and known reference materials were employed to guarantee analytical accuracy.

Using Minteq software [44] chemical speciation may be estimated from an internal Minteq thermochemical data for specified pH intervals. Establishing a reasonably constant ionic strength using the background solution chemistry [NO_3, Cl, NH_4, Ca, K, Mg, Na, SO_4, PO_4] of subsurface tile-drainage effluent from the David M. Barton Agriculture Research Center [Missouri, USA], activity coef-ficients were calculated using the Debye-Huckel equation at 25°C.

7. Results and discussion

Soils of the Sharkey series (very-fine, smectitic, thermic chromic epiaquerts) have Ap-Bssg-Bssyg horizon sequences, and soils of the Lilbourn series (coarse-loamy, mixed, superactive, nonacid, thermic aeric fluvaquents) have Ap-C horizon sequences. The Sharkey and Lilbourn soil series are composed of Holocene fluvial sediments from the ancestral Mississippi/Ohio rivers in southeastern Missouri (USA). The clayey-textured Sharkey soil series shows greater REE concentrations than the coarse-textured Lilbourn series, and both series exhibit appreciably greater than unity LREE/HREE concentration ratios. In general, the REE distri-butions obey the Oddo-Harkins rule. REE water extract concentrations are an approximate estimate of soil REE activity. As expected, the water extract concen-trations for the Sharkey and Lilbourn soil series are approximately two to three orders of magnitude smaller than the aqua regia digestion extract concentrations

	Sharkey soil series						Lilbourn soil series			
	Soil (mg kg^{-1})			Water (µg kg^{-1})			Soil (mg kg^{-1})		Water (µg kg^{-1})	
	Ap	Bssg	Bssyg	Ap	Bssg	Bssyg	Ap	C4	Ap	C4
La	24.9	24.4	24.9	43	50	41	15.5	16.8	54.5	51.1
Ce	51.1	47.6	50.8	96.3	106	86.8	29.9	34.9	554	57.6
Pr	6.4	6.1	6.2	12	15	12	3.7	4.4	13.4	14.2
Nd	25.7	24.6	24.3	51	63	49	13.8	17.0	52.8	55.9
Sm	5.0	4.9	4.7	12	15	11	2.5	3.1	11.3	13.0
Eu	1.1	1.1	1.1	3	3	2	0.5	0.7	2.4	2.9
Gd	4.9	4.8	4.6	11	14	9.6	2.1	2.8	10.8	12.0
Tb	0.7	0.7	0.6	2	2	1	0.3	0.4	1.4	1.7
Dy	3.8	3.9	3.7	7.9	10	6.9	1.57	2.12	7.3	8.4
Ho	0.7	0.7	0.7	2	2	1	0.3	0.4	1.5	1.6
Er	1.9	1.9	1.8	5	6	4	0.8	1.1	4.3	4.5
Tm	0.3	0.3	0.2	1	1	0	0.1	0.1	0.6	0.6
Yb	1.4	1.4	1.3	4	5	3	0.6	0.9	3.9	4.1
Lu	0.2	0.2	0.2	1	1	0	0	0.1	0.6	0.6

Ap is the surface horizon and Bssg, Bssyg, and C4 are subsurface horizons. [unpublished soil data from the author of this manuscript].

Table 5.
Soil rare earth element abundances for soil (mg kg^{-1}) and water extract (µg kg^{-1}).

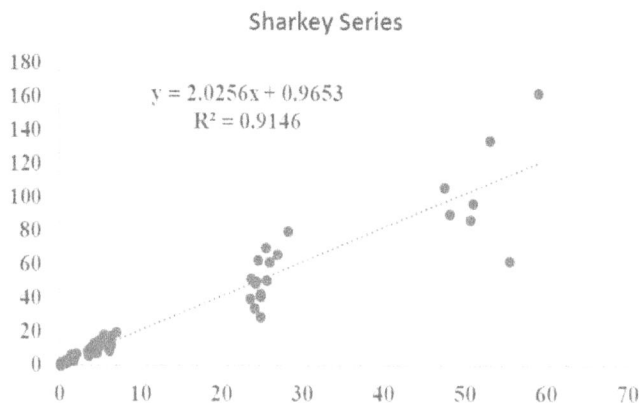

Figure 1.
The relationship between total rare earth element concentrations (x-axis with units of mg kg^{-1}) and water extractable rare earth element concentrations (y-axis with units of µg kg^{-1}) for the Sharkey soil series.

(**Table 5**). The REE distribution of the water extracts closely parallels the REE distribution of the aqua regia digestion distribution, inferring that (i) the REE release to water is influenced by the REE abundance regardless of atomic number and (ii) the water-absorbent partitioning is not strongly influenced by soil profile position (**Figure 1**).

The Kaintuck soil series in Missouri (coarse-loamy, siliceous, superactive, nonacid, mesic typic udifluvents) are very deep and well-drained floodplain soils formed from loamy alluvium and have an Ap-C horizon sequence. As with the Sharkey and Lilbourn

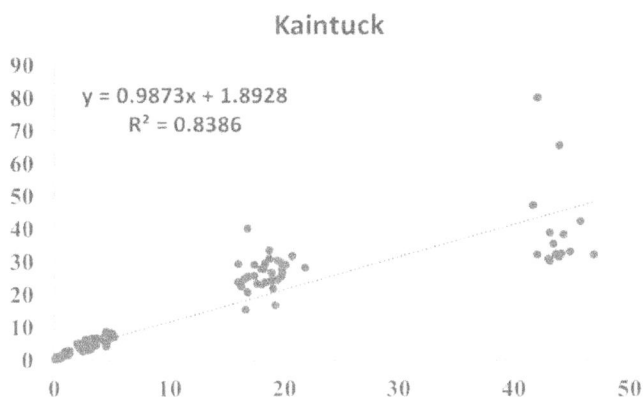

Figure 2.
The relationship between total rare earth element concentrations (x-axis with units of mg kg^{-1}) and water extractable rare earth element concentrations (y-axis with units of µg kg^{-1}) for the Kaintuck soil series.

	Ap	C1	C3	C5	C7	Ap	C1	C3	C5	C7
		Soil (mg kg^{-1})					Water (µg kg^{-1})			
La	16.6	16.8	16.8	16.0	18.2	15.5	40.2	25.5	29.3	27.9
Ce	43.1	43.9	43.9	42.0	43.0	30.3	65.7	31.5	80.2	30.9
Pr	4.5	4.5	4.6	4.4	4.7	4.3	8.6	6.3	7.3	7.5
Nd	19.2	18.7	19.7	18.3	20.1	16.9	33.6	25.6	29.7	29.2
Sm	3.2	3.7	2.8	3.3	3.2	3.4	6.6	5.3	6.2	6.2
Eu	0.6	0.6	0.6	0.6	0.6	0.7	1.5	1.2	1.4	1.4
Gd	3.0	2.9	2.8	2.7	2.9	3.2	6.0	5.0	5.7	5.7
Tb	0.3	0.3	0.4	0.3	0.3	0.5	0.9	0.8	0.9	0.9
Dy	2.5	2.5	2.3	2.1	2.2	2.6	5.1	4.1	4.7	4.8
Ho	0.4	0.4	0.4	0.4	0.4	0.5	1.0	0.8	0.9	0.9
Er	1.2	1.3	1.2	1.1	1.2	1.5	3.0	2.2	2.5	2.5
Tm	0.1	0.1	0.1	0.1	0.1	0.2	0.4	0.3	0.3	0.3
Yb	0.9	1.0	0.9	0.8	0.8	1.3	2.6	1.8	2.1	2.0
Lu	0.1	0.1	0.1	0.1	0.1	0.3	0.5	0.3	0.3	0.3

Ap is the surface horizon and C1, C3, C5, and C7 are subsurface horizons. [unpublished soil data from the author of this manuscript].

Table 6.
Soil rare earth element abundances for the Kaintuck soil series (mg kg^{-1}) and associated water extract (µg kg^{-1}).

soil series, the release of REEs to the water is a function of REE abundance, regardless of atomic number. The regression slope for the Kaintuck soil series (**Figure 2**) is smaller than the corresponding Sharkey soil series (**Figure 1**), suggesting that the binding relationships involving REE release to water are slightly different (**Table 6**).

8. REE simulations involving inorganic and organic complexation

Background electrolyte concentrations were obtained from tile-drainage water at the David M. Barton Agriculture Research Center of Southeast Missouri State University. The background total elemental concentrations (mol kg-water^{-1})

Lanthanum speciation			
Species	pH 4	pH 6	pH 8
−log (activity) (% speciation at given pH)			
La^{3+}	7.02 (92.55%)	7.22 (60.62%)	8.50 (2%)
LaCl^{2+}	9.55 (0.15%)	9.75 (0.095%)	11.10
LaSO$_4$$^+$	7.75(0.01%)	10.66 (4.13%)	11.51 (0.34%)
LaNO$_3$$^{2+}$	9.86 (0.073%)	10.06 (0.047%)	11.32
LaH$_2$PO$_4$$^{2+}$	10.66 (0.012%)	10.93	14.09
LaHCO$_3$$^{2+}$	9.14 (0.389%)	7.52 (16.32%)	8.84 (0.63%)
LaCO$_3$$^+$	11.07	7.45 (12.84%)	6.78 (57.87%)
La(CO$_3$)$_2$$^-$	—	9.85 (0.052%)	7.22 (21.22%)
LaOH^{2+}	—	10.03 (0.05%)	9.31 (0.21%)
FA$_2$-La$^+$	8.97 (0.35%)	7.74 (5.93%)	7.27 (17.74%)
Ytterbium speciation			
Species	pH 4	pH 6	pH 8
−log (activity) (% speciation at given pH)			
Yb^{3+}	10.36 (12.72%)	11.98 (0.31%)	14.31
YbCl^{2+}	13.01 (0.015%)	14.03	16.93
YbSO$_4$$^+$	11.18 (0.72%)	12.80 (0.017%)	14.91
Yb(SO$_4$)$_2$$^{-1}$	13.90	15.52	17.41
YbNO$_3$$^{2+}$	13.49	15.12	17.42
YbHPO$_4$$^+$	13.61	13.31	15.50
YbPO$_4$	15.09	12.79 (0.016%)	12.98 (0.01%)
YbHCO$_3$$^{2+}$	12.28 (0.082%)	12.09 (0.14%)	14.45
YbCO$_3$$^+$	13.33	11.14 (0.79%)	11.50 (0.33%)
Yb(CO$_3$)$_2$$^-$	18.62	12.62 (0.026%)	11.02 (0.99%)
YbOH^{2+}	13.60	13.22	13.55
FA$_2$-Yb$^+$	9.06 (86.44%)	9.00 (98.69%)	8.99 (98.66%)
Fulvic acid was set at 0.082 mg/L with estimated properties using the Stockholm Humic Model.			

Table 7.
La and Yb equilibria in a simulated natural water environment where La^{3+} or Yb^{3+} and their hydrolysis products are permitted speciation reactions with fulvic acid, carbonate, phosphate, sulfate, chloride, and nitrate complexes at three pH intervals and permitting precipitation of calcite, dolomite, and hydroxyapatite.

were (i) Ca^{2+} was 0.0032 mol kg^{-1}, (ii) CO$_3$ was 0.0079 mol kg^{-1}, (iii) Mg^{2+} was 0.0032 mol kg^{-1}, (iv) Na$^+$ was 0.0025 mol kg^{-1}, (v) NH$_4$$^+$ was 2.8 × 10^{-6} mol kg^{-1}, (vi) NO$_3$$^-$ was 0.00032 mol kg^{-1}, (vii) PO$_4$ was 10^{-4} mol kg^{-1}, (viii) Cl$^-$ was 10^{-3} mol kg^{-1}, (ix) DOC 4.16 × 10^{-6} mol kg^{-1}, (x) SO$_4$$^{2-}$ was 10^{-4} mol kg^{-1}, (xi) La^{3+} was (when simulated) 3.1 × 10^{-7} mol kg^{-1}, and (xii) Yb^{3+} (when simulated) was 1.04 × 10^{-9} mol kg^{-1}. The ionic strength was 0.0158 mol kg^{-1}. In this simulation, hydroxyapatite, dolomite, and calcite were permitted to precipitate as finite solids.

Lanthanum and ytterbium were simulated at pH 4, 6, and 8 to estimate hydrolysis, inorganic, and fulvic acid complexation. At pH 4, La^{3+} and LaHCO$_3$$^{2+}$ are the dominant species (**Table 7**), whereas fulvic acid-La complex was estimated to be present at 0.35%. At pH 8, LaCO$_3$$^+$ and La(CO$_3$)$_2$$^-$ are the dominant species, with

REE	REE^{3+}	REE-phthalic	REE(OH)$^{2+}$	REE(CO$_3$)$^+$	REE(CO$_3$)$_2^-$
		Percent of total REE			
La	0.99	18.8	0.20	48.28	31.25
Ce	0.51	—	0.31	52.49	45.82
Nd	0.22	10.90	0.19	37.67	50.93
Sm	0.12	3.97	0.23	31.39	64.24
Gd	0.14	3.83	0.27	30.29	64.90
Dy	0.06	1.22	0.21	20.10	78.31
Er	0.04	1.52	0.16	14.81	83.45
Yb	0.03	0.53	0.20	15.47	83.25

Concentration of phthalic acid (benzene-1,2-dicarboxylic acid) is 10^{-3} mol kg^{-1}. Total REE concentrations are 3.1×10^{-7} mol/kg.

Table 8.
REE complexation with phthalic acid at pH 8.3 in calcite-saturated water.

Species	Percent of species
Lanthanum	
Fulvic acid-La$^+$	5.6
LaCO$_3^+$	56.8
La(CO$_3$)$_2^-$	35.8
Cerium	
Fulvic acid-Ce$^+$	2.9
CeCO$_3^+$	51.6
Ce(CO$_3$)$_2^-$	43.9
Neodymium	
Fulvic acid-Nd$^+$	5.2
NdCO$_3^+$	40.7
Nd(CO$_3$)$_2^-$	53.6
Samarium	
Fulvic acid-Sm$^+$	7.0
SmCO$_3^+$	30.9
Sm(CO$_3$)$_2^-$	61.7
Gadolinium	
Fulvic acid-Gd$^+$	3.9
GdCO$_3^+$	30.8
Gd(CO$_3$)$_2^-$	64.3
Dysprosium	
Fulvic acid-Dy$^+$	0.8
DyCO$_3^+$	20.6
Dy(CO$_3$)$_2^-$	78.3
Erbium	
Fulvic Acid-Er$^+$	0.5
ErCO$_3^+$	15.3

Species	Percent of species
$Er(CO_3)_2^-$	83.9
Ytterbium	
Fulvic acid-Yb^+	1.3
$YbCO_3^+$	15.7
$Yb(CO_3)_2^-$	82.2

All rare earth element concentrations are initially set as 0.31×10^{-7} mol/kg. Ionic strength was estimated at 3.85×10^{-3} mol/L. Fulvic acid was set at 0.082 mg/L with estimated properties using the Stockholm Humic Model.

Table 9.
Simulation of the partitioning of selected rare earth elements between carbonate complexes and fulvic acid in stream waters in equilibrium with ordered dolomite and hydroxyapatite.

the fulvic acid-La complex being estimated to be present at 17.7%. At pH 4, Yb^{3+} and fulvic acid-Yb are the dominant species (**Table 7**), with the fulvic acid-Yb complex showing 86% of the total Yb concentration. At pH 8, the fulvic acid-Yb complex was estimated to be present at 99% of the total Yb concentration.

The REEs were simulated in the presence of phthalic acid (10^{-3} mol kg^{-1}) at pH 8.3. In this simulation dolomite, calcite, and hydroxyapatite were permitted to precipitate (**Table 8**). The REE—phthalic acid complexes as a percentage of the total REE concentration—was greatest for La and declined with increasing atomic number. The REE concentrations of $REE(CO_3)^+$ and $REE(CO_3)_2^-$ were the most extensive species, with the concentration of $REE(CO_3)^+$ declining with increasing atomic number and the concentration of $REE(CO_3)_2^-$ increasing with atomic number.

The simulation of the REE partitioning between carbonate complexes and fulvic acid in stream waters in equilibrium with ordered dolomite and hydroxyapatite was performed (**Table 9**). The fulvic acid-REE complexes generally represented less than 10% of the total REE concentration. Conversely, the concentrations of $REECO_3^+$ and $REE(CO_3)_2^-$ declined and increased, respectively, on progression with increasing atomic number.

9. Evolution of REE studies and needs

Kautenburger et al. [45] demonstrated that (i) humic acid and (ii) humic acid with partially blocked phenolic OH and COOH groups supported different complex stability constants, showing that humic acids with a high concentration of strong binding sites can be responsible for increased REE mobility because of dissolved negatively charged metal-humate complexes. Marang et al. [46] investigated the competitive behavior of Cu and Ca on Eu binding with sedimentary humic acid. Copper^{2+} and Eu^{3+} were shown to exhibit direct competition with humic acid, whereas Ca^{2+} competition was indirect and attributed to simple electrostatic interactions. Sonke [47] evaluated complexation of river, coal, and soil humic acid binding of rare earth elements. Upon progression from La to Lu, the observed increase in complex stability is consistent with lanthanide contraction and supports the premise that organic matter outcompetes carbonate complexation, even in alkaline environments, and that REE fractionation in aquatic environments is common.

Aosai et al. [48] employed nanofiltration membranes to estimate organic colloids in deep groundwaters. Ramirez-Guinart et al. [49] observed soil sorption and desorption of Sm were predicated on the Sm concentration, with dilute Sm concentrations exhibiting higher sorption and reduced desorption. Sorption of Sm was influenced by pH and soil organic matter solubility, and the soil phases of organic matter, presence of carbonates, and clay separated were important predictors of Sm mobility.

10. Future research needs

Our collective understanding of rare earth element activity in surface and groundwater requires a more fundamental examination of (i) REE partitioning within the aqueous phase, including complexation and adsorption reactions involving organic and inorganic colloids; (ii) partitioning involving REE in the aqueous phase and the surrounding solid phases constituting the river bed and aquifer skeleton; (iii) the influence of temperature, Eh (pe), pH, and ionic strength; and (iv) a greater and more accurate thermodynamic database of organic and inorganic species.

We also need a more significant database of rare earth element abundances in surface and groundwaters to gauge the extent of environmental impact and to serve as a reference for future REE environmental impact in water. Key areas of extensive groundwater and surface water across North and South America, Europe, Africa, and Asia have not received any preliminary documentation of their rare earth element composition.

Author details

Michael Aide
Southeast Missouri State University, Missouri, USA

*Address all correspondence to: mtaide@semo.edu

IntechOpen

References

[1] Lee JD. Concise Inorganic Chemistry. NY: Chapman and Hall; 1992

[2] Baes CF, Mesmer RE. The Hydrolysis of Cations. NY: John Wiley and Sons; 1976

[3] Hummel E, Berner U, Curti E, Thoenen A. Nagra/PSI Chemical Thermodynamic Data Base. Wettingen, Switzerland: Nagra; 2008

[4] Smith R, Martell A. Critical stability constants. In: Inorganic Complexes. Vol. 1 and 4. New York: Plenum Press; 1976

[5] Schijf J, Byrne RH. Stability constants for mono- and dioxalato-complexes of Y and the REE, potentially important species in groundwaters and surface freshwaters. Geochimica et Cosmochimica Acta. 2001;**65**:1037-1046

[6] Luo YR, Byrne RH. Carbonate complexation of yttrium and rare earth elements in natural waters. Geochimica et Cosmochimica Acta. 2004;**68**:691-699

[7] Cantrell KJ, Byrne RH. Rare earth element complexation by carbonate and oxalate ions. Geochimica et Cosmochimica Acta. 1987;**51**:597-605

[8] Gramaccioli CM, Diella V, Demartin F. The role of fluoride complexes in REE geochemistry and the importance of 4f electrons: Some examples in minerals. European Journal of Mineralogy. 1999;**11**:983-992

[9] Lee JH, Byrne RH. Complexation of trivalent rare earth elements (Ce, Eu, Gd, Tb, Yb) by carbonate ions. Geochimica et Cosmochimica Acta. 1993;**57**:295-302

[10] Millero FJ. Stability constants for the formation of rare earth inorganic complexes as a function of ionic strength. Geochimica et Cosmochimica Acta. 1992;**56**:3123-3132

[11] Klungness GD, Byrne RH. Comparative hydrolysis behavior of rare earth elements and yttrium: The influence of temperature and ionic strength. Polydron. 2000;**19**:99-107

[12] Aide MT. Lanthanide soil chemistry and its importance in understanding soil pathways: Mobility, plant uptake and soil health. In: Lanthanides. Rijeka, Croatia: InTech; 2018

[13] Gu ZM, Wang XR, Gu XY, Cheng J, Wang LS, Dai LM, et al. Determination of stability constants for rare earth elements and fulvic acids extracted from different soils. Talanta. 2001;**53**:1163-1170

[14] Dong MW, Li WJ, Tao ZY. Use of the ion exchange method for the determination of stability constants of trivalent metal complexes with humic and fulvic acids II. Tb^{3+}, Yb^{3+} and Gd^{3+} complexes in weakly alkaline conditions. Applied Radiation and Isotopes. 2002;**56**:967-974

[15] Pourret O, Davranche M, Gruau G, Dia A. Rare earth elements complexation with humic acid. Chemical Geology. 2007;**243**:128-141

[16] McLennan SM. Rare earth elements in sedimentary rocks: Influence of provenance and sedimentary processes. In: Lipin BR, McKay GA, editors. Geochemistry and Mineralogy of Rare Earth Elements. Reviews in Mineralogy. Vol. 21. Washington, DC: Mineralogical Society of America; 1989

[17] Kabata-Pendias A. Trace Elements in Soils and Plants. New York: CRC Press; 2001

[18] Liang T, Li K, Wang L. State of rare earth elements in different environmental components in mining areas of China. Environmental Monitoring and Assessment. 2014;**186**:1499-1513

[19] Dupre B, Viers J, Dandurand JL, Polve M, Benezeth P, Vervier P, et al. Major and trace elements associated with colloids in organic-rich river waters: Ultrafiltration of natural and spiked solutions. Chemical Geology. 1999;**160**:63-80

[20] Garcia MG, Lecomte KL, Pasquini AI, Formica SM, Depetris PJ. Sources of dissolved REE in mountainous streams draining granitic rocks, Sierras Pampeanas (Cordoba, Argentina). Geochimica et Cosmochimica Acta. 2007;**71**:5355-5368

[21] Andersson PS, Dahlqvist R, Ingri J, Gustapsson O. The isotopic composition of Nd in a boreal river: A reflection of selective weathering and colloid transport. Geochimica et Cosmochimica Acta. 2001;**65**:521-527

[22] Ingri J, Winderlund A, Land M, Gustafsson O, Andersson P, Öhlander B. Temporal variations in the fractionation of the rare earth elements in a boreal river; the role of colloidal particles. Chemical Geology. 2000;**166**:23-45

[23] Gurumurthy GP, Balakrishna K, Tripti M, Audry S, Riotte J, Braun JJ, et al. Geochemical behavior of dissolved trace elements in a monsoon-dominated tropical river basin, southwestern India. Environmental Science and Pollution Research. 2014;**21**:5098-5120

[24] Neal C. Lanthanum, cerium, praseodymium and yttrium in waters in an upland acidic and acid sensitive environment, mid-Wales. Hydrology and Earth System Sciences. 2005;**9**:645-656

[25] Leybourne MI, Johannesson KH. Rare earth elements (REE) and yttrium in stream waters, stream sediments, and Fe-Mn oxyhydroxides: Fractionation, speciation, and controls over REE + Y patterns in the surface environment. Geochimica et Cosmochimica Acta. 2008;**72**:5962-5983

[26] Ohlander B, Land M, Ingri J, Widerlund A. Mobility and transport of Nd isotopes in the vadose zone during weathering granite till in a boreal forest. Aquatic Geochemistry. 2014;**20**:1-17

[27] Dia A, Gruau G, Olivie-Lauquet G, Riou C, Molenat J, Curmi P. The distribution of rare earth elements in groundwaters: Assessing the role of source-rock composition, redox changes and colloidal particles. Geochimica et Cosmochimica Acta. 2000;**64**:4131-4151

[28] Pourret O, Gruau G, Dia A, Davranche M, Molenat J. Colloidal control on the distribution of rare earth elements in shallow groundwaters. Aquatic Geochemistry. 2010;**16**:31-59. DOI: 10.1007/s10498-009-0 or https://hal-insu.archives-ouveretes.fr/insu-00562460

[29] Coppin F, Berger G, Bauer A, Castet S, Loubet M. Sorption of lanthanides on smectite and kaolinite. Chemical Geology. 2002;**182**:57-68

[30] Chen S, Gui H. Hydrogeochemical characteristics of groundwater in the coal-bearing aquifer of the Wugou coal mine, northern Anhui Province, China. Applied Water Science. 2017;**7**:1903-1920

[31] Omonona OV, Okoghue CO. Geochemistry of rare earth elements in groundwater from different aquifers in the Gboko area, central Benue Trough, Nigeria. Environment and Earth Science. 2017;**76**:18-35

[32] Gruau G, Dia A, Olivie-Lauquet G, Davranche M, Pinay G. Controls on the distribution of rare earth elements in shallow groundwaters. Water Research. 2004;**38**:3576-3586

[33] Tang J, Johannesson KH. Speciation of rare earth elements in natural terrestrial waters: Assessing the role of dissolved organic matter from the modeling approach. Geochimica et Cosmochimica Acta. 2003;**67**:2321-2339

[34] Rabung T, Pierret MC, Bauer A, Geckeis H, Bradbury MH, Baeyens B. Sorption of Eu(III)/Cm(III) on Ca-montmorillonite and Na-illite. Part 1: Batch sorption and time-resolved laser fluorescence spectroscopy experiments. Geochimica et Cosmochimica Acta. 2005;**69**:5393-5402

[35] Dong WM, Wang XK, Bian XY, Wang AX, Du JZ, Tao ZY. Comparative study on sorption/desorption of radioeuropium on alumina, bentonite and red earth: Effects of pH, ionic strength, fulvic acid, and iron oxides in red earth. Applied Radiation and Isotopes. 2001;**54**:603-610

[36] Wu ZH, Luo J, Guo HY, Wang XR, Yang CS. Adsorption isotherms of lanthanum to soil constituents and effects of pH, EDTA and fulvic acid on adsorption of lanthanum onto goethite and humic acid. Chemical Speciation & Bioavailability. 2001;**13**:75-81

[37] Davranche M, Pourret O, Gruau G, Dia A, Le Coz-Bouhnik M. Competitive binding of REE to humic acid and manganese oxide: Impact of reaction kinetics on development of cerium anomaly and REE adsorption. Chemical Geology. 2008;**247**:154-170

[38] Davranche M, Pourret O, Gruau G, Dia A. Impact of humate complexation on the adsorption of REE onto Fe-oxyhydroxide. Journal of Colloid and Interface Science. 2004;**277**:271-279

[39] Bradbury MH, Baeyens B. Sorption of Eu on Na- and Ca-montmorillonites: Experimental investigations and modelling with cation exchange and surface complexation. Geochimica et Cosmochimica Acta. 2002;**66**:2325-2334

[40] Bradbury MH, Baeyens B, Geckeis H, Rabung T. Sorption of Eu(III)/Cm(III) on Ca-montmorillonite and Na-illite. Part 2: Surface complexation modelling. Geochimica et Cosmochimica Acta. 2005;**69**:5403-5412

[41] Sinitsyn VA, Aja SU, Kulik DA, Wood SA. Acid-base surface chemistry and sorption of some lanthanides on K$^+$ - saturated Marblehead illite: I. Results on an experimental investigation. Geochimica et Cosmochimica Acta. 2000;**64**:185-194

[42] Cteiner ZS. The influence of pH and temperature on the aqueous geochemistry of neodymium in near surface conditions. Environmental Monitoring and Assessment. 2009;**151**:279-287

[43] Pourret O, Davranche M, Gruau G, Dia A. Competition between humic acid and carbonates for rare earth elements complexation. Journal of Colloid and Interface Science. 2007;**305**:25-31

[44] Allison JD, Brown DS, Novo-Gradac KL. Minteqa2/Prodefa2, A geochemical assessment model for environmental systems: version 3.0. Environmental Research Laboratory, Office of Research and Development, U.S. Environmental Protection Agency; Athens, GA; 1991

[45] Kautenburger R, Hein C, Beck HP, and Sander JM. 2014. Influence of metal loading and humic acid functional groups on the complexation behavior of trivalent lanthanides analyzed by CE-ICP-MS. Analytica Chimica Acta 816: 50-59. [http://dx.doi.org/10.1016/j.aca.2014.01.044]

[46] Marang L, Reiller PE, Eidner S. Combining spectroscopic and potentiometric approaches to characterize competitive binding to humic substances. Environmental Science & Technology. 2008;**42**:5094-5098 [http://dx.doi.org/10.1021/es702858p]

[47] Sonke JE. Lanthanide-humic substances complexation. II. Calibration of humic ion-binding model V. Environmental Science & Technology. 2006;**40**:7481-7487 [http://dx.doi.org/10.1021/es060490g]

[48] Aosai D, Saeki D, Natsuyama H, and Iwatsuki T. Concentration and characterization of organic colloids in deep granitic groundwater using nanofiltration membranes for evaluating radionuclide transport. Colloids and Surfaces A: Physicochemical and Engineering Aspects. 2015;**485**:55-62 [http://dx.doi.org/10.1016/j.colsurfa.2015.09.012]

[49] Ramirez-Guinart O, Salaberria A, Vidal M, and Rigol A. Dependence of samarium-soil interaction on samarium concentration: Implications for environmental risk assessment. Environmental Pollution. 2018;**234**: 439- 447 [http://dx.doi.org/10.1016/j.envpol.2017.11.072]

Chapter 2

Allanite from Granitic Rocks of the Moldanubian Batholith (Central European Variscan Belt)

Miloš René

Abstract

Allanite occurs as a relative rare REE mineral in selected granitic rocks of the Moldanubian batholith. This batholith represents one of the largest plutonic bodies in the European Variscan belt. Allanite was found in the Schlieren biotite granites and diorites 1 of the oldest Weinsberg suite, in biotite granodiorites of the youngest Freistadt suite and in dykes of microgranodiorites occurred in the eastern margin of the Klenov pluton. A majority of analyzed allanites are without any magmatic zoning, only allanite grains from the diorites 1 display complicated internal zoning with variable concentrations of Fe, Ca, Th, and REE. Analyzed allanites from the Schlieren granite, diorite 1, and the "margin" variety of the Freistadt granodiorite display ferriallanite-allanite substitution with low $Fe_{ox} = (Fe^{3+}/(Fe^{3+} + Fe^{2+}))$ ratio (0.2–0.5). The analyzed allanites occurring in the microgranodiorites display slightly greater $Fe_{ox} = (Fe^{3+}/(Fe^3 + Fe^{2+}))$ ratios (0.45–0.6) and enrichment in Al (up to 2.2 apfu). All analyzed allanites are Mn-poor with its concentrations from 0.01 to 0.04 apfu. The Ce is a predominant rare earth element in all analyzed allanite grains; they are thus identified as allanite-(Ce). The highest concentrations of Ce were found in allanites from diorite 1 (0.31–0.41 apfu).

Keywords: allanite, petrology, geochemistry, cerium, bohemian massif, Moldanubian zone

1. Introduction

Allanite ($[Ca, REE]_2[Fe, Al]_3Si_3O_{12}[OH]$) is a common accessory mineral from the epidote group which occurs in intermediate granitic rocks (granodiorites, tonalites, and diorites) and their dyke equivalents (microgranodiorites and microdiorites) (e.g., [1–3]). Although its modal abundance in these rocks is low, allanite is a major residence site for LREE. It is related to epidote by coupled substitution:

$$REE^{3+} + Fe^{2+} = Ca^{2+} + Fe^{3+} \tag{1}$$

and to clinozoisite by

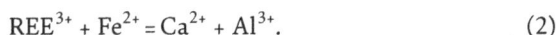

$$REE^{3+} + Fe^{2+} = Ca^{2+} + Al^{3+}. \tag{2}$$

This manuscript concentrates on mineralogy and chemical composition of allanite which occurs as a relatively rare accessory mineral in some intermediate granitic rocks of the Moldanubian batholith of the Bohemian Massif. The Moldanubian batholith represents a large plutonic body in the Bohemian Massif composed of biotite granodiorites, granites, and two-mica granites together with some younger dykes (aplites, pegmatites, felsic granites, and microgranodiorites to microdiorites) [4, 5].

2. Geological setting

The Moldanubian batholith forms one of the plutonic complexes within the Central European Variscan belt, covering 10,000 km^2 [5] (**Figure 1**). In detail, the Moldanubian batholith is built by multiple plutons, predominantly composed of granitic to granodioritic rocks with either S- or transitional I/S-type character [5–7]. All these granitic rocks can be classified into three main suites. These three suites are represented as (1) coarse-grained, porphyritic I- to I/S-type biotite granites to

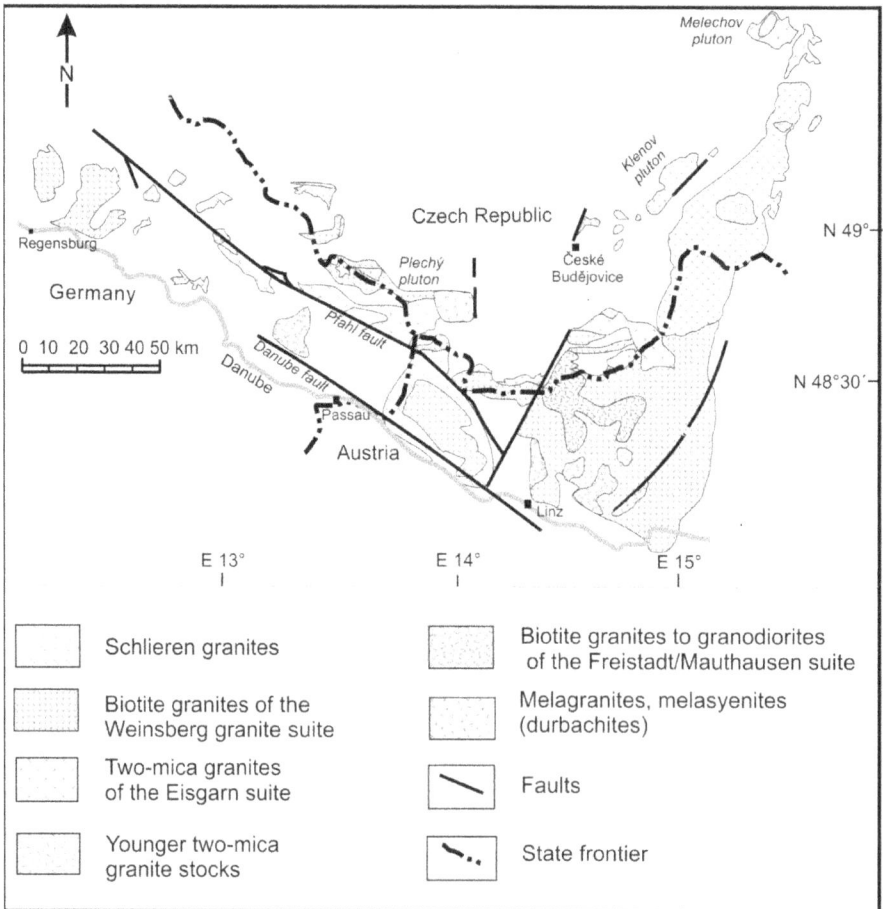

Figure 1.
Geological map of the Moldanubian batholith (after [5], modified by the author).

granodiorites of the Weinsberg suite, (2) medium grained, partly porphyritic two-mica S-type granites of the Eisgarn suite, and (3) fine- to medium-grained I/S-type biotite granites to granodiorites of the Freistadt/Mauthausen suite [5, 6, 8].

A significant part of the Weinsberg suite is in situ evolved Schlieren granite, which occurs in the Upper Mühlviertel area (Austria) and attached area of the Bavaria (Germany). Diffuse and irregular contacts, transitional rock varieties, and intrusion of one granite to the other indicate that the Schlieren and Weinsberg granites coexisted as magmas; thus, they are of the same age [9]. However, in the past, the Schlieren granite was originally mapped and described as "coarse grained gneiss" [10]. With intrusion of the Weinsberg granite suite in the Bavarian and Austrian part of the Moldanubian batholith are also connected intrusions of diorite stocks (diorite 1) [11].

Two petrographic varieties were identified in the main body of the Freistadt suite in the Austrian Mühlviertel, the coarse-grained "marginal variety", and medium-grained "central variety" [12]. Allanite, however, occurs only in granodiorites of the "marginal variety".

The granitic rocks of the Moldanubian batholith are in some cases intruded by dykes of microdiorites, microgranodiorites, granite and melasyenite porphyries, and stock of highly fractionated two-mica and muscovite granites [11, 13–19].

3. Sampling and methods

Allanite was more commonly found in the Schlieren granite of the Weinsberg suite. As a relatively rare accessory mineral, allanite occurs also in diorites connected with granodiorites of the Weinsberg suite, in granodiorites of the Freistadt/Mauthausen suite and in microgranodiorites occurring on the eastern margin of the Klenov pluton.

Allanite together with selected rock-forming minerals (plagioclase, biotite) was analyzed in polished thin sections. The back-scattered electron (BSE) images were acquired to study the internal structure of individual allanite grains. Element abundances of Al, Ca, Ce, Dy, Er, Eu, F, Fe, Gd, Ho, La, Lu, Mg, Mn, Na, Nd, P, Pb, Pr, Sc, Si, Sm, Sr, Tb, Th, Ti, Tm, U, Y, and Yb were determined using a CAMECA SX-100 electron microprobe operated in wavelength-dispersive mode. The concentrations of these elements were determined using an accelerating voltage and a beam current of 15 kV and 20 nA, respectively, with a beam diameter of 2–5 μm. The following standards, X-ray lines, and crystals (in parentheses) were used: AlK_α—sanidine (TAP), CaK_α—fluorapatite (PET), CeL_α—$CePO_4$ (PET), DyL_α—$DyPO_4$ (LiF), ErL_α—$ErPO_4$ (PET), EuL_β—$EuPO_4$ (LIF), FeK_α—almandine (LiF), GdL_β—$GdPO_4$ (LiF), HoL_β—$HoPO_4$ LiF), LaL_α—$LaPO_4$ (PET), LuM_β—LuAg (TAP), MgK_α—spessartine (LIF), NdL_β—$NdPO_4$ (LIF), PK_α—fluorapatite (PET), PbM_α—vanadinite (PET), PrL_α—$PrPO_4$ (LIF), SrL_α—$SrSO_4$ (TAP), ScK_α—ScP_5O_{14} (PET), SiK_α—sanidine (TAP), SmL_β—$SmPO_4$ (LIF), TbL_α—$TbPO_4$ (LIF), ThM_β—$CaTh(PO_4)_2$ (PET), TiK_α—anatas (PET), TmL_α—$TmPO_4$ (LiF), UM_β—metallic U (PET), and YL_α—YPO_4 (PET). Intra-REE overlaps were partially resolved using L_α and L_β lines. Empirically determined coincidences were applied after analysis: ThM_α on the PbM_α line and ThM_γ on the UM_β line. The raw data were converted into concentrations using appropriate PAP-matrix corrections [20]. The detection limits were approximately 400 pm for Y, 180–1700 ppm for REE, and 800–1000 ppm for U and Th. The plot (REE + Y + Th + Mn + Sr) vs. Al proposed by Petrík et al. [21] was used for estimation of the $Fe_{ox} = Fe^{3+}/(Fe^{3+} + Fe^{2+})$ ratio by electron microprobe analyzed allanite.

4. Petrography

The Schlieren granites of the Weinsberg suite are represented by biotite granites consisting of plagioclase (An_{20-40}) (32–50 vol.%), K-feldspar (7–37 vol.%), quartz (18–34 vol.%), and biotite (annite, Fe/Fe + Mg = 0.53–0.55, Al^{4+} = 2.10–2.13, and Ti = 0.23–0.42 atoms per formula unit (apfu)), (6–32 vol.%). Amphibole was also frequently present (up to 5 vol.%). Accessory minerals are represented by apatite, zircon, ilmenite, magnetite, titanite, and allanite.

The biotite diorites (diorite 1) of the Weinsberg suite consist of plagioclase (An_{37-39}), (50–53 vol.%), biotite (annite, Fe/Fe + Mg = 0.65–0.66, Al^{4+} = 2.22–2.93, Ti = 0.37–0.45 apfu), (15–20 vol.%), K-feldspar (8–12 vol.%), amphibole (Mg-hornblende), (3–10 vol.%), and pyroxene (Fe-augite), (1–5 vol.%). Accessory minerals are represented by ilmenite, apatite, zircon, titanite, allanite, and thorite.

The biotite granodiorites of the "marginal variety" of the Freistadt suite consist of plagioclase (An_{25-37}), (32–68 vol.%), quartz (12–32 vol.%), K-feldspar (3–27 vol.%), biotite (annite, Fe/Fe + Mg = 0.44–0.62, Al^{4+} = 2.09–2.28, and Ti = 0.30–0.50 apfu), (6–17 vol.%) and muscovite (0–1 vol.%). Accessory minerals are represented by apatite, zircon, ilmenite, titanite, monazite, and allanite.

The microgranodiorites from the eastern margin of the Klenov pluton consist of plagioclase (An_{25-54}), K-feldspar, quartz, biotite (annite, Fe/Fe + Mg = 0.60–0.68, Al^{4+} = 1.68–2.30, and Ti = 0.13–0.47 apfu), pyroxene (Fe-augite), and amphibole (ferro-actinolite to Mg-hornblende). Accessory minerals are represented by ilmenite, titanite, apatite, rutile, zircon, and rare allanite.

5. Mineralogy and mineral chemistry of allanite

Allanite in these rock types occurs as a rare accessory mineral. It forms in these rocks relatively bigger grains (300–500 μm) and usually occurs on grain boundaries of biotite and plagioclase. Electron microprobe data show that the chemical composition of the epidote-group minerals in analyzed granitic rocks of the Moldanubian batholith varies greatly (**Table 1**). Studied allanites often exhibit irregular alteration, usually along their grain rims without any zoning of unaltered parties. In the BSE images, highly altered allanite parties on their rims are dark (**Figure 2A**). These highly altered parties are enriched in Si, Ti, and Th and depleted in Ca, Fe, Mn, La, and Ce. The altered allanite parties also display lower total analytical sum, which could indicate their hydration. In some other cases, irregular bright parties of BSE in altered allanite grains were found. These bright parties are enriched in Fe and depleted in Si, Ti, Ca, and Th (**Figure 2B**).

Analyzed epidote-group minerals without visible alteration contain 30.9–36.1 wt.% SiO_2, 10.1–17.8 wt.% CaO, 8.8–15.1 wt.% FeO, and 13.0–24.4 wt.% REE_2O_3. The magmatic zoning observed in some analyzed allanite grains (**Figure 2C–E**) seems to be caused by variations in Fe, Ca, Th, and REE contents and $Fe^{3+}/(Fe^{3+} Fe^{2+})$ ratio. Allanites from the Schlieren granites and Freistadt suite are relatively Al-poor (Al = 1.3–1.8 atoms per formula unit, apfu) and display variable $Fe_{ox} = (Fe^{3+}/(Fe^{3+} + Fe^{2+}))$ ratio (0.2–0.5). Allanites from the diorite 1 are enriched in Al (1.8–1.9 apfu). Distinctly greater Al enrichment occurs in allanites from microgranodiorites (up to 2.2 apfu). These allanites also display higher $Fe_{ox} = (Fe^{3+}/(Fe^{3+} Fe^{2+}))$ ratio without any zoning (**Figure 2F**). All analyzed allanites are Mn-poor with its concentrations from 0.01 to 0.04 apfu.

Sample	1714-16	1714-19	1724-11	1633-7	1633-12	532-30
Suite	Weinsberg	Weinsberg	Weinsberg	Freistadt	Freistadt	Dykes
Variety wt.%	Schlieren granite	Schlieren granite	Diorite 1	Margin	Margin	Micro-granodiorite
SiO_2	35.97	34.74	32.01	31.47	31.67	32.36
TiO_2	1.28	1.29	1.22	1.06	1.20	1.17
Al_2O_3	14.88	14.38	15.78	13.45	14.06	14.69
FeO	8.66	9.33	13.43	15.10	15.20	14.94
MnO	0.43	0.52	0.30	0.41	0.47	0.21
MgO	0.92	1.10	0.91	1.52	1.16	0.24
CaO	11.04	10.89	10.62	10.78	11.34	12.48
La_2O_3	5.15	4.89	6.67	7.40	6.46	5.37
Ce_2O_3	8.89	9.47	12.29	11.48	10.95	8.96
Pr_2O_3	0.61	0.78	1.25	1.09	1.09	0.87
Nd_2O_3	1.91	2.46	3.59	2.85	2.87	2.92
Sm_2O_3	0.07	0.21	0.29	0.08	0.26	0.52
Gd_2O_3	0.00	0.23	0.22	0.16	0.12	0.22
Dy_2O_3	0.00	0.00	0.01	0.17	0.19	0.08
ThO_2	2.80	2.33	0.20	1.12	1.10	0.10
UO_2	0.04	0.05	0.01	0.02	0.01	b.d.l.
Total	92.65	92.67	98.80	98.16	98.15	95.13
apfu, O = 12.5						
Si	3.42	3.35	3.04	3.06	3.05	3.12
Ti	0.09	0.09	0.09	0.08	0.09	0.09
Al	1.66	1.63	1.77	1.54	1.60	1.67
Fe^{2+}	0.69	0.75	1.07	1.23	1.23	1.20
Mn	0.04	0.04	0.02	0.03	0.04	0.02
Mg	0.13	0.16	0.13	0.22	0.17	0.03
Ca	1.12	1.13	1.08	1.12	1.17	1.29
La	0.18	0.17	0.23	0.25	0.23	0.19
Ce	0.31	0.33	0.43	0.43	0.39	0.32
Pr	0.02	0.03	0.04	0.04	0.04	0.03
Nd	0.07	0.09	0.12	0.11	0.10	0.10
Sm	0.00	0.01	0.01	0.01	0.01	0.02
Gd	0.00	0.01	0.01	0.00	0.00	0.01
Dy	0.00	0.00	0.00	0.01	0.01	0.00
Th	0.06	0.05	0.00	0.03	0.02	0.00
U	0.00	0.00	0.00	0.00	0.00	0.00

b.d.l.—below detection limit and apfu—atoms per formula unit.

Table 1.
Selected representative microprobe analyses of allanite.

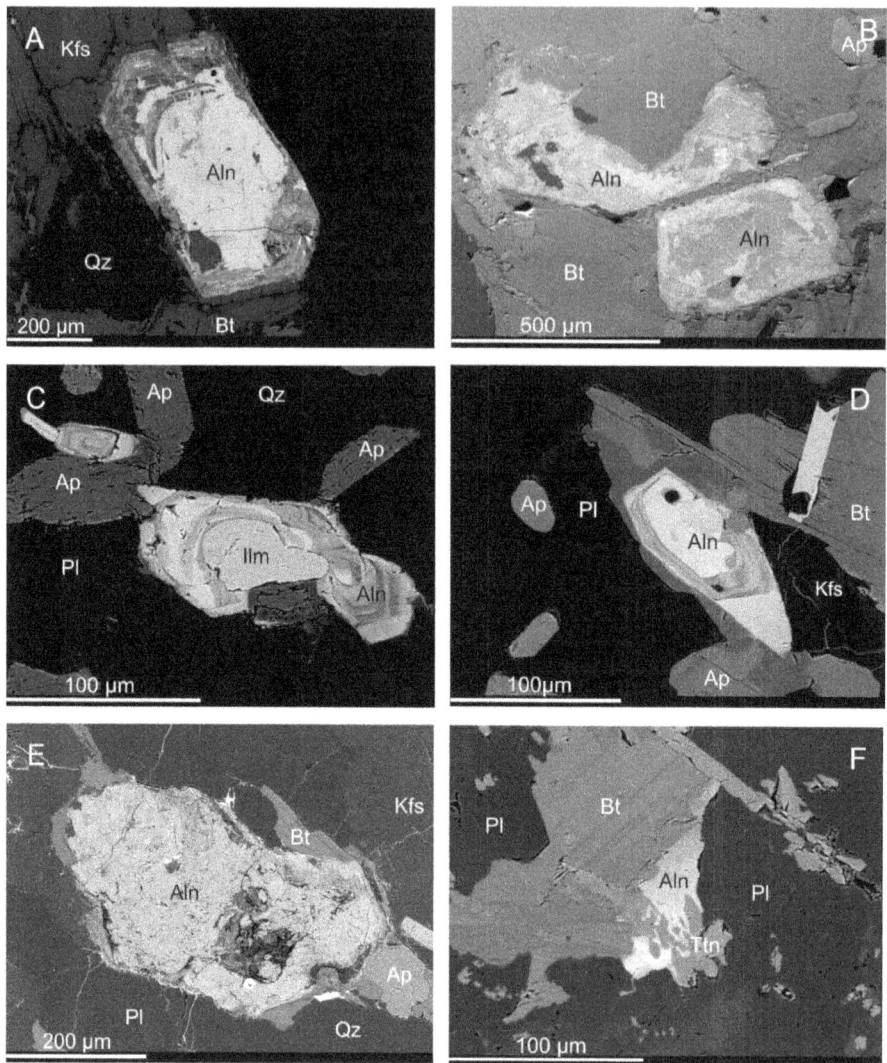

Figure 2.
BSE images of allanite, A—partly altered allanite grain from biotite granodiorite of the Freistadt suite, B—highly altered allanite grain form the Schlieren granite, C and D—zoned allanite grains from the diorite 1, E—allanite grain from the Schlieren granite, and F—allanite grain from microgranodiorite. Aln—allanite, Ap—apatite, Bt—biotite, Ilm—ilmenite, Kfs—K-feldspar, Qz—quartz, Pl—plagioclase, and Ttn—titanite.

The majority of analyzed allanites represent substitution between ferriallanite and allanite. Only allanites from the microgranodiorites display substitution between allanite and clinozoisite (**Figure 3**).

All analyzed allanites display variable concentrations of REE with preference of Ce over La. The Ce is predominant in all analyzed allanite grains over other REE studied here; they are thus identified as allanite-(Ce). The highest concentrations of Ce were found in allanites from diorite 1 (0.31–0.41 apfu). The lowest concentrations of Ce display allanites from the youngest microgranodiorite dykes (0.14–0.32 apfu).

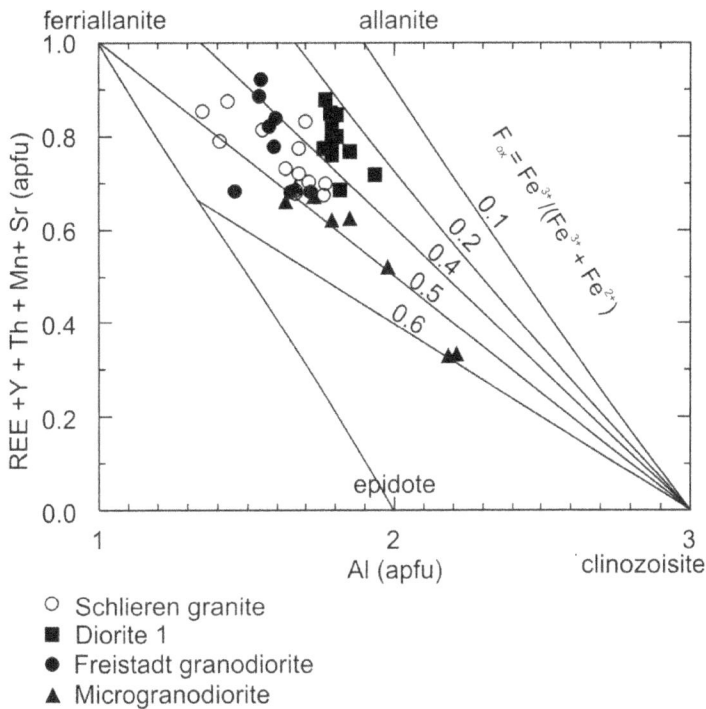

Figure 3.
Plot of REE + Y + Th + Mn + Sr. (apfu) vs. Al (apfu) with isolines of the ratio $Fe_{ox} = Fe^{3+}/(Fe^3 + Fe^{2+})$ after Petrík et al. [21].

6. Discussion

For allanite, two main substitutions occur, namely the epidote-allanite and the allanite-ferriallanite substitutions [2, 21]. For analyzed allanites from the Weinsberg and Freistadt suites, the allanite-ferriallanite substitution is significant. Similar substitution was found by Petrík et al. [21] in allanites from the I-type granitic rocks of the Sihla tonalite suite in the Western Carpathians. Some highly altered allanite grains which were found in the Schlieren granite (**Figure 2B**) exhibit irregular zonation, which is very similar with the "mushroom-shaped areas" described by Poitrasson [22] from anorogenic granites of Corsica (southeast France). However, in the case of altered allanites from the Schlieren granite, the altered parties are depleted in Si, Ti, Ca, and Th, but enriched in Fe. Some other allanite alteration was found on allanite rims that occurred in the allanite from the Freistadt granodiorite (**Figure 2A**). In this case, the altered allanite rim is enriched in Si, Ti, and Th. Similar enrichment of Th was also found in altered alanites from anorogenic granites of Corsica (southeast France) and in allanites from the Casto granite of Idaho (USA) [22, 23]. Alterations of allanite which were found in allanites from the Schlieren granite and Freistadt granodiorite could be very probably explained by later late- and post-Variscan alteration of the Moldanubian batholith, which was connected with Pb-Zn and U-mineralization, which occurs in this region.

The allanite grains display in some cases three types of zoning, as revealed in BSE images: (1) oscillatory zoning [1, 24], (2) normal growth-induced magmatic zoning [2, 22], and (3) complicated internal zoning consisting of a patchwork of domains variable in brightness [21]. In allanite grains from diorite 1, complicated internal zoning was found (**Figure 2C** and **D**).

The allanite-clinozoisite substitution that is significant for allanite from microgranodiorites occurring in the eastern margin of the Klenov pluton was also found in allanites from epidote-bearing tonalites in the Bell Island pluton, Canada [25].

7. Conclusions

Allanite occurs in some intermediate to basic igneous rocks of the Moldanubian batholith. It was found in the oldest Schlieren granites and diorite 1 of the Weinsberg suite, in the youngest granodiorites of the Freistadt/Masuthausen suite, and in selected dykes composed of microgranodiorites in the eastern margin of the Klenov pluton. Analyzed allanites from the Schlieren granite, diorite 1, and the "margin" variety of the Freistadt granodiorite display ferriallanite-allanite substitution and a low $Fe_{ox} = (Fe^{3+}/(Fe^{3+} + Fe^{2+}))$ ratio (0.2–0.5). The analyzed allanites occurring in microgranodiorites display partly higher $Fe_{ox} = (Fe^{3+}/(Fe^{3+} + Fe^{2+}))$ ratio (0.45–0.6) and enrichment in Al (up to 2.2 apfu).

The allanites from the Schlieren granite and Freistadt granodiorite display in some cases variable alteration which is coupled with different behaviors of Si, Fe, Ti, and Th. This alteration is very probably connected with late- and/or post-Variscan hydrothermal alteration of these granitic rocks.

Acknowledgements

This study was carried out thanks to the support of the long-term conceptual development research organization RVO 67985891 and the project of the Ministry of Education, Youth and Sports (ME10083). I am also grateful to P. Gadas and R. Škoda from the Institute of Masaryk University for technical assistance by electron microprobe analyses of selected minerals (allanite, plagioclase, biotite, amphibole, and pyroxene). I am also grateful to Michael Aide for numerous comments and recommendations that helped to improve this paper.

Conflict of interest

The author declares no conflict of interest.

Author details

Miloš René
Institute of Rock Structure and Mechanics, Academy of Sciences of the Czech Republic, Prague, Czech Republic

*Address all correspondence to: rene@irsm.cas.cz

IntechOpen

References

[1] Deer WA, Howie RA, Zussman J. Rock-Forming Minerals, vol. 1B Disilicates and Ringsilicates. 2nd ed. Harlow: Longham; 1986. p. 629

[2] Giére R, Sorensen SS. Allanite and other REE-rich epidote-group minerals. Reviews in Mineralogy and Geochemistry. 2004;**56**:431-493. DOI: 10.2138/gsrmg.56.1.431

[3] Armbruster T, Bonazzi P, Akasaka M, Bermanec V, Chopin C, Giéré R, et al. Recommended nomenclature of epidote-group minerals. European Journal of Mineralogy. 2006;**18**:551-567. DOI: 10.1127/0935-1221/2006/0018-0551

[4] Finger F, Gerdes A, René M, Riegler G. The Saxo-Danubian granite belt: Magmatic response to post-collisional delamination of mantle lithosphere berlow the south-western sector of the bohemian massif (Variscan orogeny). Geologica Carpathica. 2009;**60**:205-212

[5] Verner K, René M, Žák J, Janoušek V. A brief introduction in the geology of the Moldanubian batholith. In: Janoušek V, Žák J, editors. Eurogranites 2015: Variscan Plutons of the Bohemian Massif. Post-Conference Field Trip Following the 26th IUGG General Assembly in Prague. Prague: Czech Geological Survey; 2015. pp. 103-109

[6] Gerdes A, Wörner G, Henk A. Post-collisional granite generation and HT/LP metamorphism by radiogenic heating: The Variscan south bohemian batholith. Journal of the Geological Society of London. 2000;**157**:577-587. DOI: 10.1144/jgs.157.3.577

[7] Breiter K. Geochemical classification of Variscan granitoids in the Moldanubicum (Czech Republic, Austria). Abhandlungen der Geologischen Bundesanstalt. 2010;**65**:19-25

[8] Vellmer C, Wedepohl KH. Geochemical characterization and origin of granitoids from the south bohemian batholith in Lower Austria. Contributions to Mineralogy and Petrology. 1994;**118**:13-32. DOI: 10.1007/BF00310608

[9] Finger F, Clemens J. Migmatization and "secondary" granitic magmas: Effects of emplacement and crystallization of "primary" granitoids in southern bohemian massif, Austria. Contributions to Mineralogy and Petrology. 1995;**120**:311-326. DOI: 10.1007/BF00306510

[10] Fuchs G. Zur Altersgliederung des Moldanubikums Oberösterreichs. Verhandlungen der Geologische Bundesanstalt. 1962:96-117

[11] Fuchs G, Thiele O. Erläuterungen zur Übersichtskarte des Kristallins in westlichen Mühlviertel und im Sauwald, Oberösterreich. Wien: Geologische Bundesanstalt; 1968. p. 96

[12] Klob H. Der Freistädter Granodiorit im österreichischen Moldanubikums. Verhandlungen der Geologische Bundesanstalt. 1971;**1**:98-142

[13] Němec D. Lamprophyrische und lamproide Ganggesteine in Südteil der Böhmisch-Mährischen Anhöhe (ČSSR). Tschermaks Mineralogische Und Petrographische Mitteilungen. 1970;**14**:235-284. DOI: 10.1007/BF01081341

[14] Vrána S, Bendl J, Buzek F. Pyroxene microgranodiorite dykes from the Ševětín structure, Czech Republic: Mineralogical, chemical and isotopic indication of a possible impact melt origin. Journal of the Czech Geological Society. 1993;**38**:129-148

[15] Breiter K, Scharbert S. Latest intrusions of the Eisgarn pluton (South

Bohemia—Northern Waldviertel). Jahrbuch der Geologischen Bundesanstalt. 1998;**141**:25-37

[16] Košler J, Kelley SP, Vrána S. ^{40}Ar/^{39}Ar hornblende dating of a microgranodiorite dyke: Implications for early Permian extension in the Moldanubian zone of the Bohemian Massif. International Journal of Earth Sciences. 2001;**90**:379-385. DOI: 10.1007/s00531000154

[17] René M. Ti-rich granodiorite porphyries from the northeastern margin of the Klenov massif (Moldanubian zone of the bohemian massif). Acta Montana Series A. 2003;**23**:77-85

[18] Harlov DE, Procházka V, Förster HJ, Matějka D. Origin of monazite-xenotime-zircon-fluorapatite assemblages in the peraluminous Melechov granite massif, Czech Republic. Mineralogy and Petrology. 2008;**94**:9-26. DOI: 10.1007/s0710-008-0003-8

[19] Žáček V, Škoda R, Sulovský P. U-Th-rich zircon, thorite and allanite-(Ce) as main carriers of radioactivity in the highly radioactive ultrapotassic melasyenite porphyry from the Šumava mts, Moldanubian zone, Czech Republic. Journal of Geosciences. 2009;**54**:343-354. DOI: 10.3190/jgeosci.053

[20] Pouchou JL, Pichoir F. PAP (φ-ρ–Z) procedure for improved quantitative microanalysis. In: Armstrong JT, editor. Microbeam Analysis. San Francisco: San Francisco Press; 1985. pp. 104-106

[21] Petrík I, Broska I, Lipka J, Siman P. Granitoid allanite-(Ce): Substitution relations, redox conditions and REE distributions (on an example of I-type granitoids, Western Carpathians, Slovakia). Geologica Carpathica. 1995;**46**:79-94

[22] Poitrasson F. In situ investigations of allanite hydrothermal alteration: Examples from calc-alkaline and anorogenic granites of Corsica (Southeast France). Contributions to Mineralogy and Petrology. 2002;**142**:485-500. DOI: 10.1007/s004100100303

[23] Wood SA, Ricktetts A. Allanite-(Ce) from the Eocene Casto granite, Idaho: Response to hydrothermal alteration. The Canadian Mineralogist. 2000;**38**:81-100. DOI: 10.2113/gscanmin.38.181

[24] Dahlquist JA. REE fractionation by accessory minerals in epidote-bearing metaluminous granitoids from the sierras Pampeanas, Argentina. Mineralogical Magazine. 2001;**65**:463-475. DOI: 10.1180/002646101750377506

[25] Beard JS, Sorensen SS, Gieré R. REE zoning in allanite related to changing partition coefficients during crystallization: Implications for REE behaviour in an epidote-bearing tonalite. Mineralogical Magazine. 2006;**70**:419-435. DOI: 10.1180/0026461067040337

Section 2

Rare Earth Applications

An Efficient Route for Synthesis of Macrocyclic Gadolinium Complexes and Their Role in Medical Applications

Rasha E. El-Mekawy

Abstract

The coordination science of gadolinium has been broadly examined in the ongoing years in light of the fact that the subsequent buildings can be advantageously utilized as a helpful devices in numerous fields going from the logical science, hydrometallurgy, science and medication. It is accounted for that the combination of the gadolinium complexes and concentrate their spectroscopic properties utilizing infrared spectroscopy (IR), mass spectroscopy (MS), electron paramagnetic reverberation (EPR) and dc magnetic susceptibility techniques. MRI contrast agents have become a basic piece of present modern magnetic resonance imaging. It has been discovered that the expansion of complexity specialists as a rule improves affectability as well as explicitness of inward body structures.

Keywords: gadolinium, polycrystalline gadolinium aluminum perovskite, gold-coated gadolinium nanocrystals, *gadolinium (III) Trifluoromethanesulfonate*

1. Introduction

Gadolinium is a substance component. Its ground state electronic setup is $[Xe]4f^7 5d6s^2$. Monazite and bastnaesite are the chief gadolinium minerals where gadolinium happen together with different individuals from the uncommon earth components or the lanthanides. It very well may be isolated from the other uncommon earths by particle trade or dissolvable extraction systems. Gadolinia, the oxide of gadolinium was first separated from the mineral gadolinite by Jean-Charles-Galissard de Marignac in 1880. (Gadolinite is named after the completion scientific expert Johan Gadolin). In 1886 Paul-Émile Lecoq de Boisbaudran autonomously isolated the oxide of gadolinium from Carl Mosander's "yttria" (sullied yttrium oxide). Gadolinium is a silver-white, pliable and bendable metal. Gadolinium metal is ferromagnetic just beneath room temperature. Gadolinium science is ruled by the trivalent gadolinium (III) particle, Gd^{3+}. This particle structures ionic bonds with ligands containing an oxygen or nitrogen giver structure. The ground state electronic setup of Gd^{3+} is $[Xe]4f^7$. In spectroscopic investigation there are no ingestion groups in the unmistakable locale of the electromagnetic range and gadolinium mixes are dry. Gadolinium (III) chelates are utilized as differentiation reagents in attractive reverberation imaging (MRI).

Because of the high attractive snapshot of the paramagnetic Gd^{3+} particle (with its seven unpaired electrons), the unwinding time of water atoms in the closeness of Gd^{3+} particles is significantly diminished and signal power is along these lines upgraded. X-ray is a medicinal demonstrative method that relies upon the proton atomic attractive reverberation signal from water in its making of a proton thickness map. Gadolinium is a noteworthy part of X-beam phosphors, for example, $Gd_2O_2S:Tb^{3+}$, inside which it weakens the dynamic producer (Tb^{3+}) to maintain a strategic distance from fixation extinguishing. Since gadolinium viably assimilates neutrons, this component has discovered some utilization in control poles for atomic reactors [1, 2].

2. Chemistry

2.1 Hydrothermal synthesis and characterization of polycrystalline gadolinium aluminum perovskite (GdAlO, GAP)

Gadolinium aluminum perovskite (GdAIO, GAP) is a promising high temperature earthenware material, known for its wide application in phosphors. Polycrystalline gadolinium aluminum perovskites were orchestrated utilizing an antecedent of co-hasten gel of GdAIO by utilizing aqueous supercritical liquid system under strain and temperature running from 150 to 3200 MPa and 600 to 700°C, separately. The came about results of GAP were considered utilizing the portrayal systems, for example, powder X-beam diffraction investigation (**Figure 1**), in infrared spectroscopy, filtering electron microscopy (**Figure 2**) and vitality dispersive examination of X-beam (EDX) (**Figure 3**). The X-beam diffraction example coordinated well with the revealed orthorhombic GAP pattern (JCPDs-46-0395) [3–5].

2.2 Synthesis of square gadolinium-oxide nanoplates

The gadolinium-oxide nanocrystals were combined by arrangement stage deterioration of gadolinium - acetic acid derivation antecedents within the sight of

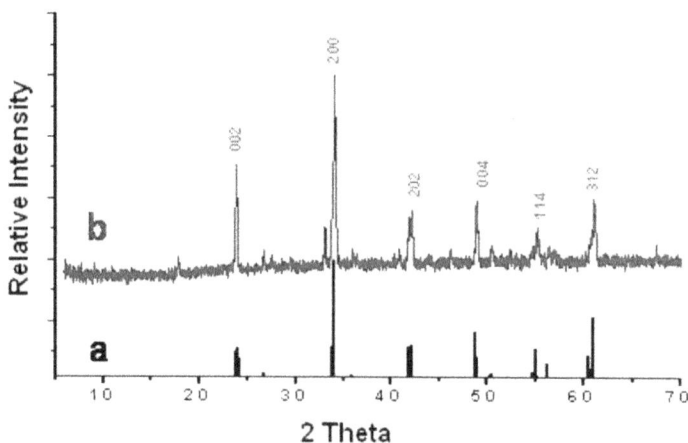

Figure 1.
XRD pattern of (a) JCPDS = 46–0395, (b) synthesized GdAlO.

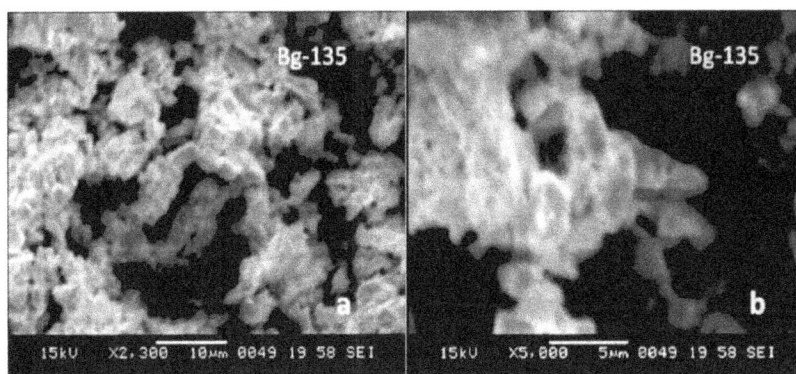

Figure 2.
EM images of GdAlO₃.

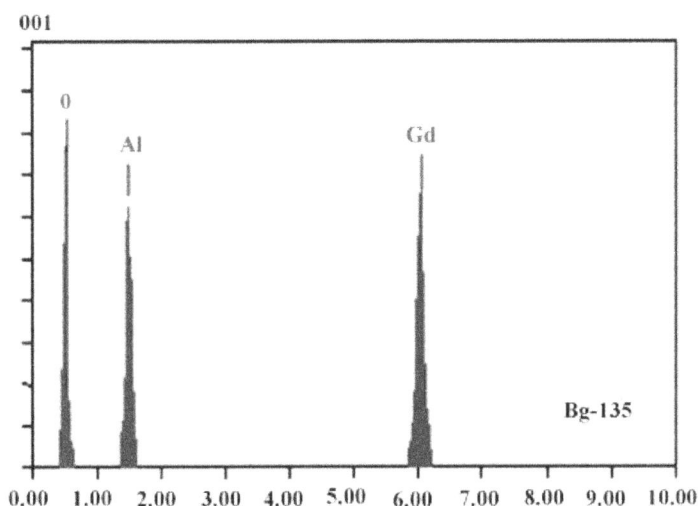

Figure 3.
EDAX of GdAlO crystal.

both organizing and noncoordinating solvents. In a run of the mill analyze, gadolinium acetic acid derivation hydrate (0.75 mmol, from Aldrich) was broken down in an answer that contained oleylamine (1.7 ml), oleic corrosive (1 ml) and octadecene (2.7 ml) at 100°C with lively blending under vacuum (~20 mTorr). Under Ar stream, the subsequent arrangement was warmed to 320°C over around 5 min. and afterward the arrangement was cooled to room temperature after 1 hr. The nanocrystals were accelerated from the response arrangement by including a blend of hexane and $(CH_3)_2CO$ (1:4) and dried under an Ar stream. The as-arranged nanocrystals are dispersible in nonpolar natural solvents, for example, toluene and hexane [6]. X-ray diffraction demonstrated that the nanocrystals comprise of crystalline Gd_2O_3. The wide-edge XRD example of nanocrystals demonstrates the trademark crests Gd_2O_3 precious stone stage which are widened due to the limited crystalline area size (**Figure 4**) [7].

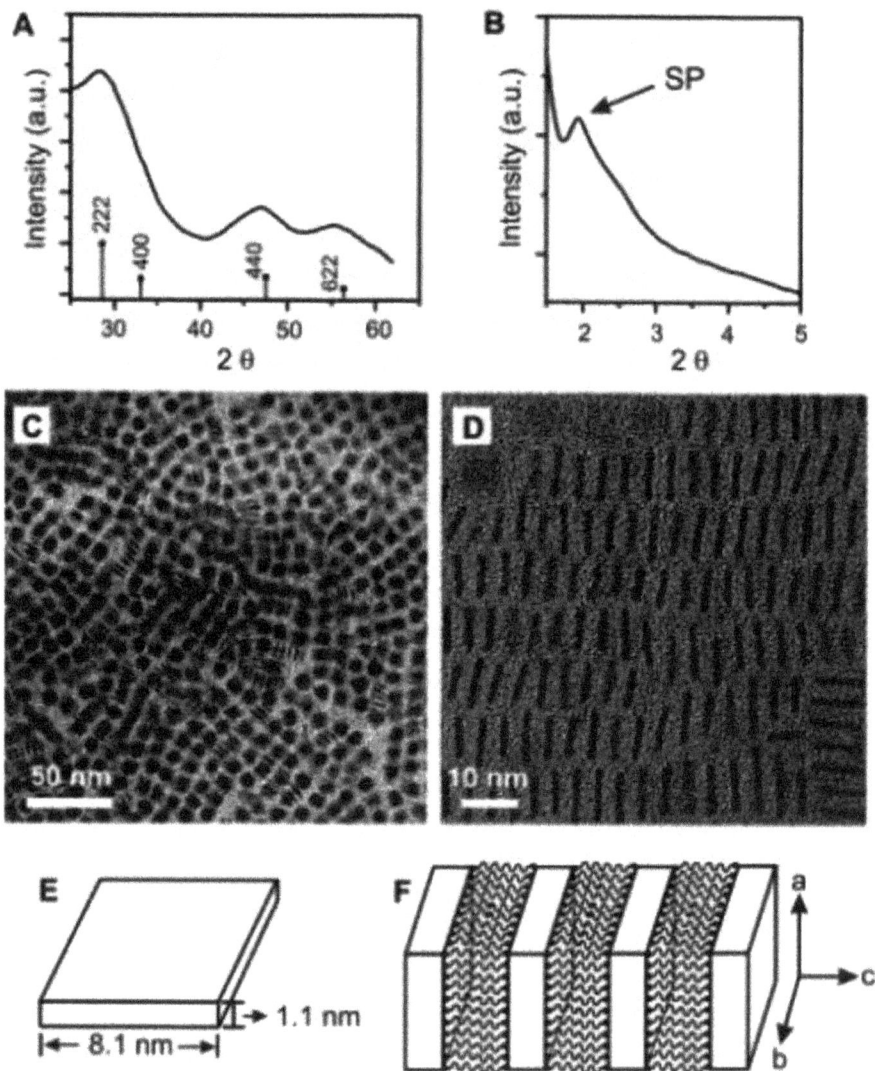

Figure 4.
(A) Wide-angle XRD. The standard diffraction peak positions of bulk cubic Gd_2O_3 are indicated. (B) Small-angle XRD. (C and D) TEM images of Gd_2O_3 nanoplates. (E and F) proposed model for the nanoplates and assembly of nanoplate stacks, respectively. The c-axis of cubic Gd_2O_3 crystals is assigned as the thickness direction of the nanoplates.

Transmission electron microscopy demonstrates that the Gd_2O_3 nanocrystals are in reality square as opposed to 3D squares. The edge length of each nanoplate is 8.1 nm with a standard deviation of 6% (**Figure 5**).

2.3 New gadolinium(III) complexes with simple organic acids (oxalic, glycolic and malic acid)

The arrangement of dreary gadolinium buildings (x, y, z), between x gadolinium particles, y ligands and z protons, of some natural acids, has been contemplated in watery arrangement. In this work we present the aftereffects of examinations on the association of the gadolinium particle [Gd(III)] with basic carboxylic acids, for

Figure 5.
TEM image of the superlattice of Gd O nanoplates. The insertis an electron-diffraction pattern taken in this area.

example, oxalic corrosive, glycolic corrosive and malic corrosive, in weaken watery arrangement with pH esteems somewhere in the range of 5.50 and 7.50. The acquired gadolinium buildings with oxalate, glycolate and malate particle are dry and haven't any ingestion band in UV-unmistakable, in this sense, the backhanded photometry concentrates used to recognize the major di-atomic and tri-atomic edifices. This method enabled us to ascertain the creations and the secure qualities constants of these major edifices in arrangement and the solidness steady relies upon acridity, the structures of the last buildings were dictated by methods for IR and Raman spectroscopies [8, 9].

2.4 Optimized routes for the preparation of gadolinium carbonate and oxide nano-particles and exploring their photocatalytic activity

A sequence of organized precipitation test was conducted based on the Taguchi robust design so as to evaluate the best conditions for the preparation of Gd (CO) nano-particles in the absence of common additives like surfactants, templates or catalysts, indicating that the dimensions of the product nano- particles can be manipulated merely through altering the parameters affecting the reaction. These parameters include the concentrations of Gd(III) and carbonate ions as well as the reactor temperature. The optimal reaction conditions led to the production of $Gd_2(CO_3)_3$ nano-particles of 36 nm in average diameter, which were evaluated by scanning electron microscopy (SEM), fourier transform infrared spectroscopy (FT-IR), thermogravimetric-differential thermal analysis, and UV–*Vis* spectrophotometry. $Gd_2(CO_3)_3$ was further calcinated at 700°C to decompose into spherical $Gd_2 O_3$ nano-particles with average diameters below 25 nm, the formation of which was established by SEM, X-ray diffraction (XRD), and FT-IR techniques. In order to obtain the band gap energies of the fabricated carbon-ate and oxide nano-products, they were characterized by UV–*Vis* diffuse reflectance spectroscopy (DRS). Besides, the photocatalytic behaviors of the nano-products in degradation of methyl orange as a pollution of water were explored, and the results exhibited the efficacy of both products in eliminating of the organic pollutant [10].

2.5 Alkalide reduction: synthesis and characterization of gold nanoparticles and gold-coated gadolinium nanocrystals

As of late, air, moisture and acid stable gold-covered gadolinium (Gd@Au) nanoparticles were set up by alkalide decrease. The union brought about center shell nanocrystals with a tight size appropriation. The Gd center could make the nano-crystals brilliant sub-atomic MRI differentiate specialists and give various alterna-tives to treatment of tumors, including 157Gd neutron catch treatment, photon initiation treatment, synchrotron stereotactic radiotherapy, and 159Gd radionuclide treatment. The Au shell anticipates Gd center from oxidation renders them stable even at low pH, which conceivably averts filtering and bio-inconsistency. T1 and T2 relaxivities demonstrate that Gd@Au nanocrystals are an extremely encouraging potential T1 MRI differentiate specialist [11].

2.6 Synthesis of a carborane gadolinium – DTPA complex for boron neutron capture therapy

Hydrolysis of the ethyl esters in 1 was done with LiOH in fluid methanol pur-sued by treatment with weakened hydrochloric corrosive (1N) to bear the cost of

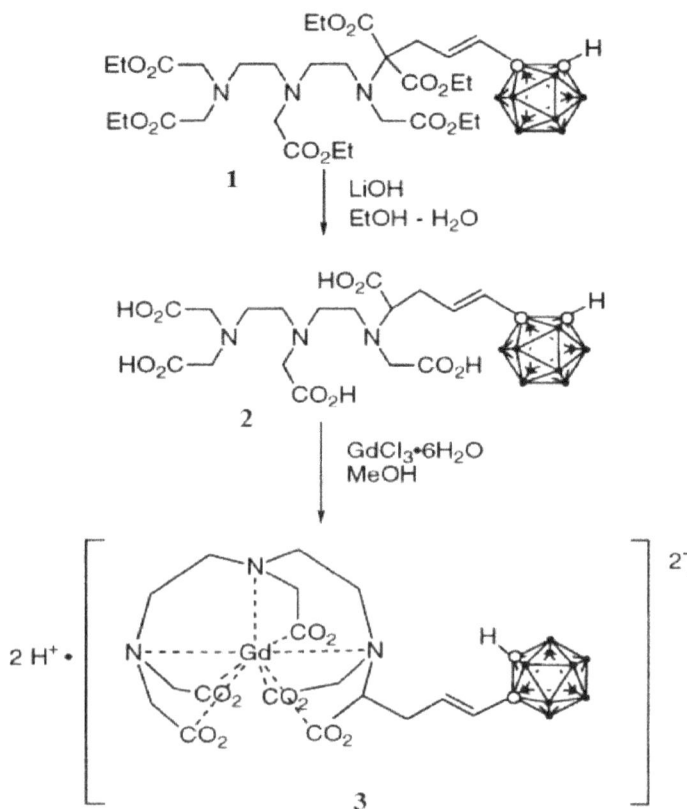

Figure 6.
Formation of the desired Gd – Carborane complex 3 in quantitative yield.

the relating pentaacid 2 in 68% yield. Treatment of the carborane containing DTPA subordinate 2 with gadolinium(III) chloride hexahydrate gave the ideal Gd – DTPA carborane complex 3 in quantitative yield (**Figure 6**) [12].

2.7 Single-crystal-to-single-crystal anion exchange in a GadoliniumMOF: incorporation of POMs and [AuCl]

The exemplification of useful atoms inside permeable coordination polymers (otherwise called metal-natural structures, MOFs) has happened to incredible enthusiasm for ongoing years at the field of multifunctional materials. In this article, we present an investigation of the impacts of size and charge in the anion trade procedure of a Gd based MOF, including atomic species like polyoxometalates '(POMs) and $[AuCl_4]^-$. This post-manufactured modification has been described by IR, EDAX, and single precious stone diffraction, which have given unequivocal proof of the area of the anion atoms in the structure [13] (**Figures 7** and **8**).

2.8 Gadolinium(III) Trifluoromethanesulfonate

Alternate Name: gadolinium(III) triflate.
Physical Data: anhydrous, $d = 7.07$ g cm^{-3}.
Solubility: soluble in water, ethanol, THF, acetone, acetonitrile, and other polar organic solvents.
Form Supplied in: commercially available as an anhydrous, white to off-white powder.

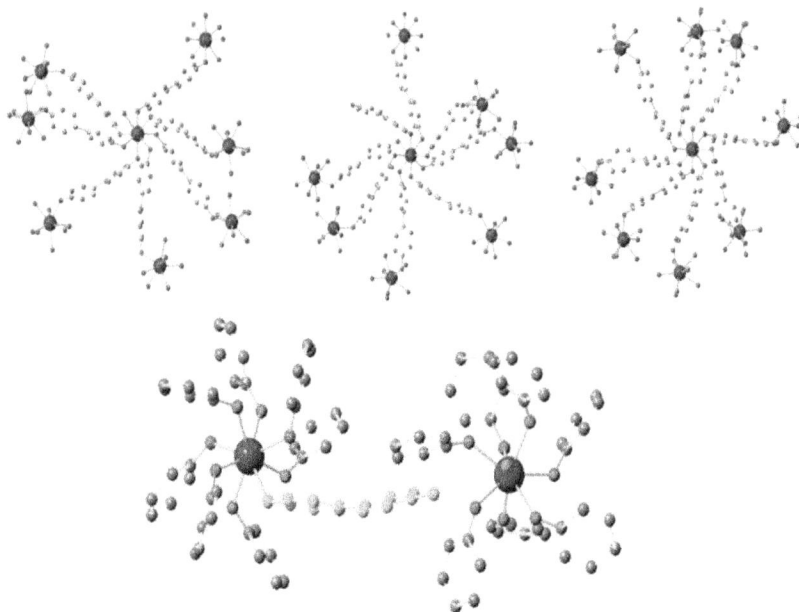

Figure 7.
(Top) view of the three crystallographically independent Gd centers in 1–AuCl₄: Gd1, Gd2 and Gd3. Gd1 has a coordinated water molecule (in addition to 7 bridging bipyNO ligands), Gd2 has a terminal bipyNO ligand (in addition to 7 bridging bipyNO ligands), and Gd3 is coordinated to 8 bipyNO ligands that serve as bridges between Gd centers. (Bottom) close view of the coordination environment of Gd1 and Gd2, highlighting the presence of a terminal bipyNO ligand (in green) and a coordinated water molecule (in pink).

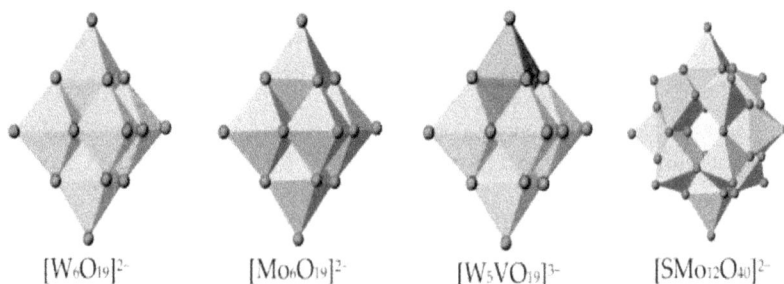

$[W_6O_{19}]^{2-}$ $[Mo_6O_{19}]^{2-}$ $[W_5VO_{19}]^{3-}$ $[SMo_{12}O_{40}]^{2-}$

Figure 8.
Structural units of the different polyoxometalates (POMs) used for anion exchange studies.

Preparative Method: the reagent is prepared by heating gadolinium oxide (Gd_2O_3) with triflic acid in water (1/1, v/v) at 100°C for 2 h. The hydrate thus prepared is extensively heated under vacuum (200°C/0.5 mmHg for 40 h) to give the anhydrous salt.

Purification: the reagent is typically used in anhydrous form as prepared above.

Handling, Storage and Precautions: skin, eye, and respiratory tract irritant. The anhydrous salt is air and water stable, hygroscopic, and should be stored in a tightly closed container in a dry and well-ventilated place. Gadolinium triflate is reported to be nonexplosive and nonflammable [14].

3. Applications

3.1 Gadolinium(III) chloride: a novel and an efficient reagent for the synthesis of homoallylic alcohols

Carbonyl compounds efficiently undergo nucleophilic addition reactions with allylstannanes in the presence of $GdCl_3.H_2O$ in acetonitrile under extremely mild reaction conditions to give the corresponding homoallylic alcohols in excellent yields and with high chemoselectivity (**Figures 9–11**) [15].

3.2 Microwave-assisted catalytic acetylation of alcohols by gold-nanoparticle–supported gadolinium complex

A gold nanoparticle (AuNP)– upheld gadolinium complex (RS-Au-L-Gd) impetus was set up through basic chelation of GdCl3 to the surface-bound spacer, 1,4,7-tris(carboxymethyl)-10-(11-mercaptoundecyl)-1,4,7,10-tetraazacyclododecane (HSDO3A). This AuNP-bolstered Gd complex was observed to be a profoundly viable impetus for the acetylation of different alcohols and phenol within the sight of acidic anhydride. With a stacking of 0.4 mol% of RS-Au-L-Gd, the practically complete change can be accomplished in 60 s under microwave light conditions. This half and half impetus was air stable, water solvent, dissolvable in numerous natural media, and precipitable. It very well may be promptly reused in excess of multiple times with no noteworthy loss of its synergist action [16].

3.3 Gadolinium (III) chloride catalyzed facile synthesis of 2-substituted benzimidazoles under solvents-free conditions

A progression of 2-substituted benzimidazoles have been set up from o-diamines and 1,3-dicarbonyl mixes utilizing Gadolinium chloride as an impetus

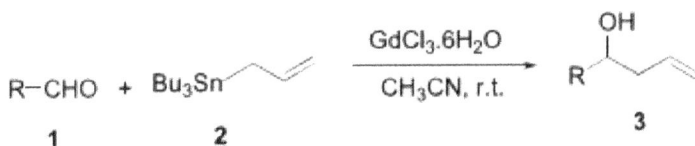

Figure 9.
Mechanism of synthesis of homollylicalcohol.

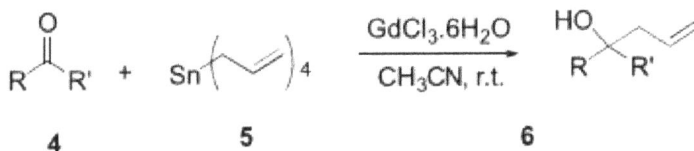

Figure 10.
Synthesis of homollylic alcohol from aldehydes.

Figure 11.
Synthesis of homollylic alcohol from ketones.

under dissolvable free condition in a decent yields. Gadolinium chloride has been exhibited as a mellow and effective impetus [17] (**Figure 12**).

3.4 In efficient gadolinium metallocene-based catalyst for the synthesis of isoprene rubber with perfect 1,4-cis microstructure and marked reactivity difference between lanthanide metallocenes toward dienes as probed by butadiene—isoprene copolymerization catalysis

An efficient gadolinium metallocene-based catalyst for the synthesis of isoprene rubber with perfect 1,4-Cis microstructure and marked reactivity difference between lanthanide metallocene toward dienes as probed by butadiene-isoprene copolymerization catalysis was studied. The large difference in catalytic reactivity effected by various lanthanide metals have been investigated by means of butadiene/isoprene copolymerization. It was observed that only the Gd-metallocene complex, which was the smallest and takes isoprene most reluctantly in the copolymerization with butadiene, can catalyze the homopolymerization of isoprene efficiently. The copolymerization reactions were carried out with varying initial monomer feed ratios at room temperature [18].

3.5 Highly dispersed ultra-small Pd nanoparticles on gadolinium hydroxide nanorods for efficient hydrogenation reaction

Heterogeneous synergist hydrogenation responses are vital to the petrochemical business and fine compound blend. In this, we present the primary case of gadolinium hydroxide [Gd(OH)₃] nanorods as a help for stacking ultra-little Pd nanoparticles for hydrogenation responses. Gd(OH)₃ has an enormous number of hydroxyl bunches superficially, which go about as a perfect help for good scattering of Pd nanoparticles.

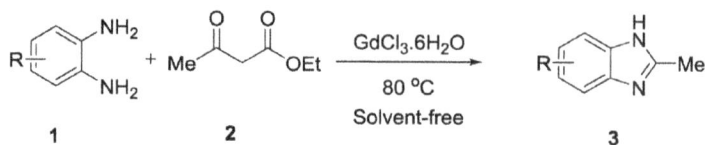

Figure 12.
General procedure for synthesis of 2-substituted benzimidazoles.

$Gd(OH)_3$ nanorods are set up by aqueous treatment, and $Pd/Gd(OH)_3$ impetus with a low stacking of 0.95 wt % Pd is acquired by photochemical testimony. The synergist hydrogenation of p-nitrophenol (4-NP) to p-aminophenol (4-AP) and styrene to ethylbenzene is executed as a model response. The acquired $Pd/Gd(OH)_3$ impetus shows amazing action when contrasted with other detailed heterogeneous impetuses. The rate consistent of 4-NP decrease is estimated to be $0.047 \, s^{-1}$ and the $Pd/Gd(OH)_3$ nanocatalyst demonstrates no stamped loss of action even after 10 back to back cycles. Also, the hydrogenation of styrene to ethylbenzene over $Pd/Gd(OH)_3$ nanorods shows a turnover recurrence (TOF) as high as $6159 \, h^{-1}$ with 100% selectivity. Besides, the impetus can be recuperated by centrifugation and reused for up to 5 back to back cycles without clear loss of movement. The outcomes show that $Gd(OH)_3$ nanorods go about as an advertiser to upgrade the reactant action by giving a synergistic impact from the solid metal help communication and the huge surface region for high scattering of little estimated Pd nanoparticles advanced with hydroxyl bunches superficially. The elite of $Pd/Gd(OH)_3$ in heterogeneous catalysis offers another, effective and effortless methodology to investigate other metal hydroxides or oxides as backings for natural changes (**Figure 13**) [19].

3.6 Gadolinium based metal: organic framework as an efficient and heterogeneous catalyst to activate epoxides for cycloaddition of CO_2 and alcoholysis

Improvement of heterogeneous impetuses for the cycloaddition of CO_2 with epoxides to get ready cyclic carbonates is a hotly debated issue in the field of CO2 change. Thus, an uncommon earth-metal gadolinium-based metal–natural system (Gd-MOF) was blended from $GdCl_3$ and pyromellitic dianhydride in N, N-dimethylformamide, which was portrayed by powder X-beam diffraction (XRD), filtering electron microscopy (SEM), transmission electron microscopy (TEM), N_2 adsorption–desorption, and Fourier change infrared (FT-IR) spectroscopy. The combined Gd-MOF could be utilized as heterogeneous impetus for the cycloaddition of CO_2 with epoxides within the sight of quaternary ammonium salts, and Gd-MOF/n-Bu_4NBr demonstrated the best execution for the cycloaddition of CO_2 with different epoxides to frame the relating cyclic carbonates because of the incredible synergetic impact. Moreover, the readied Gd-MOF could be utilized as heterogeneous impetus for alcoholysis of different epoxides to frame β-alkoxy alcohols successfully [20].

3.7 The gadolinium (Gd^{3+}) and tin (Sn^{4+}) Co-doped $BiFeO_3$ nanoparticles as new solar light active photocatalyst

The procedure of photocatalysis is engaging tremendous intrigue inspired by the extraordinary guarantee of tending to current vitality and natural issues through changing over sun oriented light straightforwardly into substance vitality. Be that as it may, an effective sun oriented vitality collecting for photocatalysis remains a basic test. Here, we revealed another full sun oriented range driven photocatalyst by co-doping of Gd^{3+} and Sn^{4+} into An and B-destinations of $BiFeO_3$ at the same time. The co-doping of Gd^{3+} and Sn^{4+} assumed a key job in hampering the recombination of electron-opening sets

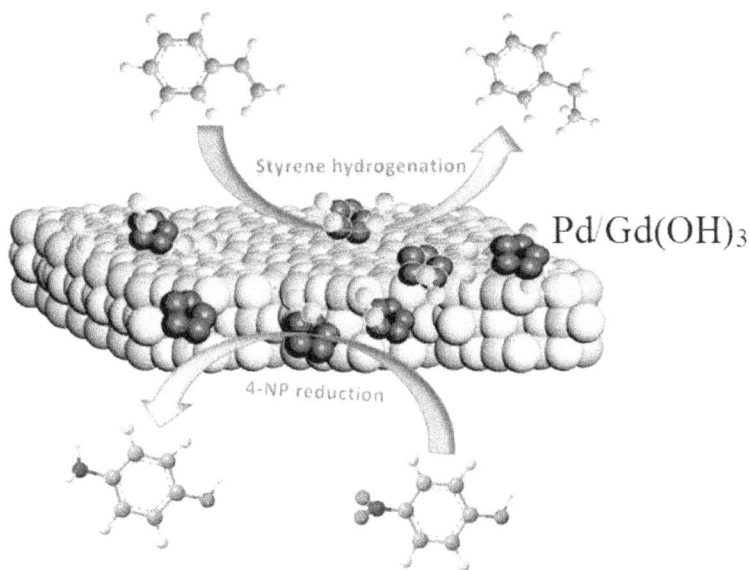

Figure 13.
Loading ultra-small Pd nanoparticles on gadolinium hydroxide nanorods.

and moved the band-hole of $BiFeO_3$ from 2.10 eV to 2.03 eV. The Brunauer-Emmett-Teller (BET) estimation affirmed that the co-doping of Gd^{3+} and Sn^{4+} into $BiFeO_3$ expanded the surface region and porosity, and in this way the photocatalytic movement of the $Bi0.90Gd0.10Fe0.95Sn0.05O_3$ framework was fundamentally improved. The work proposed another photocatalyst that could corrupt different natural colors like Congo red, Methylene blue, and Methyl violet under illumination with various light wave-lengths and gave direction for planning progressively productive photocatalysts [21].

3.8 Mechanical downsizing of a gadolinium(III)-based metal: organic framework for anticancer drug delivery

A Gd(III)-based permeable metal–natural system (MOF), Gd-pDBI, has been blended utilizing fluorescent linker pDBI (pDBI = (1,4-bis(5-carboxy-1H-benzimid-azole-2-yl)benzene)), bringing about a three-dimensional interpenetrated structure with an one-dimensional open channel (1.9 × 1.2 nm) loaded up with hydrogen-bonded water gatherings. Gd-pDBI shows high warm steadiness, porosity, amazing water solid-ness, alongside organic-solvent and mellow corrosive and base soundness with main-tenance of crystallinity. Gd-pDBI was changed to the nanoscale system (ca. 140 nm) by mechanical pounding to yield MG-Gd-pDBI with phenomenal water dispersibility (>90 min), keeping up its porosity and crystallinity. In vitro and in vivo investigations on MG-Gd-pDBI uncovered its low blood poisonous quality and most noteworthy medication stacking (12 wt%) of anticancer medication doxorubicin in MOFs answered to date with pH-responsive cancer-cell-specific medication discharge [22].

3.9 Role of gadolinium in magnetic resonance imaging (MRI)

3.9.1 What is gadolinium differentiate medium?

Gadolinium differentiate media (some of the time called a MRI differenti-ate media, specialists or "colors") are synthetic substances utilized in attractive reverberation imaging (MRI) checks. At the point when infused into the body, gadolinium differentiate medium upgrades and improves the nature of the MRI pictures (or pictures). This permits the radiologist (a pro specialist prepared to look at the pictures and give a composed report to your primary care physician or authority) to all the more precisely report on how your body is functioning and whether there is any sickness or variation from the norm present.

Gadolinium differentiate media comprise of complex particles; game plans of iotas held together by substance bonds. The synthetic bonds are made between a gadolinium particle and a bearer atom (a chelating operator). A chelating specialist averts the lethality of gadolinium while keeping up its difference properties. Various brands of gadolinium differentiate medium utilize distinctive chelating atoms. The complexity medium is infused intravenously (into a vein) as a major aspect of a MRI examine, and wiped out from the body through the kidneys [23].

3.9.2 Dose gadolinium in magnetic resonance imaging

The portion of gadolinium differentiate for babies and kids ought to be equivalent to that given to grown-ups on a for every kilogram premise. This suggestion considers two contending variables deciding the viability of complexity organization: natural half-life and volume of conveyance.

Since babies and youthful newborn children have essentially lower glomerular filtration and renal freedom rates than do more seasoned kids and grown-ups, the biologic half-existence of gadolinium-based MR differentiate operators is drawn out. In a full-term infant the half-life is 6.5 h; it might be longer than 9 h in untimely newborn children. By 2 months of age, the baby half-life achieves the grown-up estimation of 1.5 h.

This drawn out half-life in babies and youthful newborn children brings about persevering upgrade of ordinary structures for as long as a few hours following infusion. The delayed half-life in this manner gives an expanded window of time for performing imaging in these patients. For instance, a quieted newborn child who stirs during difference imbuement might be expelled from the imager, re-calmed, and re-imaged inside 1–2 h without the requirement for infusion of extra gadolinium. On the other hand, if just a postcontrast study is wanted, the newborn child might be quieted and given the differentiation imbuement while still in the neonatal consideration unit, and afterward may experience nonurgent MR imaging (**Figure 14**) [24].

Renal discharge rates aside, neonates have double the volume of extracellular liquid than grown-ups have in extent to their body loads. Hence neonates and youthful newborn children who get gadolinium differentiate on a portion for every kilogram premise will have blood gadolinium groupings of just a single a large portion of that in grown-ups after equilibration. This reality contends against utilizing a lower portion for every kilogram in newborn children than in grown-ups, despite the fact that the serum half-life is drawn out. Clinical experience by our gathering and others has exhibited that the grown-up portion of gadolinium differentiate (0.1 mmol/kg for most extracellular operators) is additionally fitting in babies and youngsters [24].

3.9.3 Are gadolinium contrast agents safe?

Gadolinium contrast agents are extremely safe. However, some patients with an allergy to such agents should consult with their doctor before a gadolinium contrast agent is used.

More recently, it has been shown that MRI can detect tiny amounts of the gadolinium in the brains of patients who have received many previous doses of gadolinium. The Food and Drug Administration has been investigating this effect since 2015. To date, no symptoms or diseases are linked to gadolinium deposition in the brain, despite

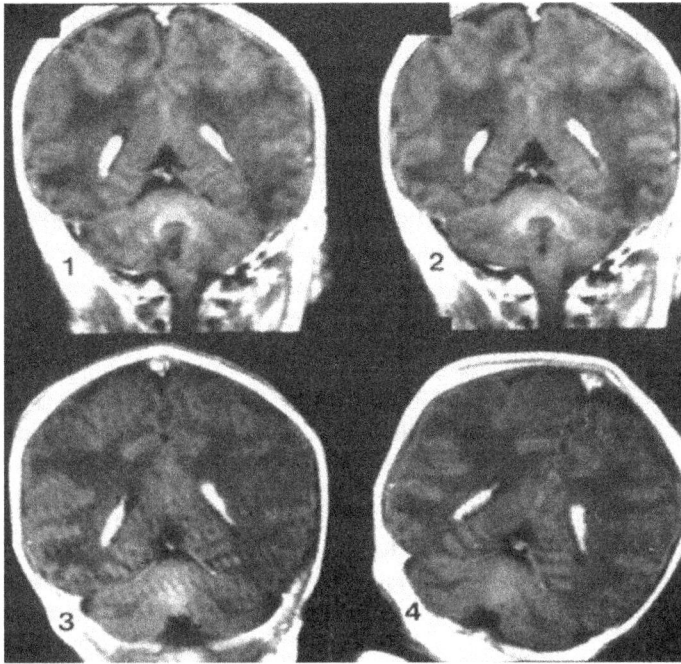

Figure 14.
Persistent difference upgrade in an untimely neonate seen at 1, 2, 3, and 4 h after infusion. The delayed half-life (4–9 h) of Gd in neonates and youthful newborn children gives an extended postcontrast imaging window.

hundreds of millions of doses administered since 1988. There continues to be research in this area to better understand this phenomenon and its possible consequences. However, to date, there are no known side effects related to this observation [25].

4. Spectroscopy of gadolinium

4.1 Spectroscopic and magnetic properties of a gadolinium macrobicyclic complex

Because of Schiff base buildup, gadolinium cryptate has been blended and researched by infrared (IR) spectroscopy and electron paramagnetic reverberation (EPR) method. Correlation of IR groups in a ligand and the gadolinium complex affirmed the arrangement of the gadolinium cryptate complex. IR and thermogravimetry-differential warm examination (TG-DTA) investigations demonstrate the nearness of two water particles in the inward circle of the complex. Mass spectroscopy examinations affirmed a monometallic substitution of the Gd^{3+} particle into the macrobicycle ligand. EPR spectra of the complex have been enrolled in the 3–300 K temperature go. Every range has been recreated utilizing the EPR–NMR PC program and the estimations of the turn Hamiltonian parameters at every temperature have been determined. The warm reliance of the turn Hamiltonian parameters has been explored. The temperature reliance of the incorporated force of the EPR range uncovered the attractive connections in the turn arrangement of this compound. No long-extend attractive request has been identified in the 3–300 K go, however a solid antiferromagnetic association in the high-temperature run, over 160 K, has been watched [26].

4.2 Magnetic and spectroscopic properties of gadolinium tripodal Schiff base complex

Gadolinium(III) tripodal Schiff base (tris(((5-chlorosalicylidene)amino)ethyl) amine) complex has been acquired and examined by infrared spectroscopy (IR), attractive vulnerability, and electron paramagnetic reverberation (EPR) strategies. Examination of IR groups in ligand and gado-linium complex confirmed the development of the gadolinium complex and permitted to propose its structure. Both electron ionization and electron splash sub-atomic spectroscopy spectra confirmed the [1:1] extent of a ligand to metal in gadolinium tripodal Schiff base complex example. IR spectroscopy and TGeDTA prohibited the nearness of water particle in the metal coordination circle. X-beam powder examination applying Fullprof PC program has demonstrated that the explored test was monophase with the monoclinic symmetry of the unit cell having the grid costants: a ¼ 10.028(4) A, b ¼ 13.282(5) A, c ¼ 21.20(1) An and b ¼ 101.58(4). Space bunch P21/c, Z ¼ 4. EPR spectra of the complex have been enrolled in the 4–300 K temperature run. Every range has been fitted utilizing EPR-NMR PC program and the estimations of the turn Hamiltonian parameters at every temperature have been determined. Temperature reliance of the coordinated power of the EPR range permitted uncovering the attractive communications in the turn arrangement of this compound. Examination of the temperature reliance of dc attractive helplessness (c) and EPR powerlessness (c˙)demonstrated significant contrasts between these amounts because of the nearness of brief clus-EPR ters with a non-attractive ground state [27].

Acknowledgements

I direct my thanks to all sides that learn me the chemistry science.

Conflict of interest

No conflict of interest to be declares.

Author details

Rasha E. El-Mekawy[1,2]

1 Department of Petrochemicals, Egyptian Petroleum Research Institute, Cairo, Egypt

2 Department of Chemistry, Faculty of Applied Science, Umm Al-Qura University, Makkah, Saudi Arabia

*Address all correspondence to: rashachemistry1@yahoo.com

IntechOpen

References

[1] Bünzli J-CG, Choppin GR, editors. Lanthanide Probes in Life, Chemical and Earth Sciences. New York: Elsevier; 1989

[2] Kaltsoyannis N, Scott P. The f Elements. New York: Oxford University Press; 1999

[3] Girish HN, Basavalingu B, Shao G-Q, Sajan CP, Verma SK. Materials Science - Poland. 2015;**33**(2):301-305

[4] Harada Y, Uekawa N, Kojima T, Kakegawa K. Journal of the European Ceramic Society. 2009;**29**:24192426

[5] Izauskaite S, Riechlova V, Enartaviciene G, Beganskiene A, Pinkas J, Kareiva A. Journal of Material Science - Poland. 2007;**25**:755-764

[6] Charles Cao Y. Synthesis of square gadolinium-oxide nanoplates. Journal of the American Chemical Society. 2004;**126**:7456-7457

[7] Alivisatos AP. Nature Biotechnology. 2004;**22**:47-54

[8] RIRI M, Hor M, Kamal O, Eljaddi T, Benjjar A, Hlaïbi M. Journal of Materials and Environmental Science. 2011;**2**(3):303-308

[9] Doyon D. Elsevier Masson. 2004;**3**:39-47

[10] Seied MP, Mehdi R-N MRG, Meisam SK, Norouzi P. Desalination and water treatment. 2017;**74**:316-325

[11] Ming Z, Graduate Student Department of Chemistry George Washington University

[12] Hisao N, Jianping C, Hiroyuki N, Masaru F, Yoshinori Y. Journal of Organometallic Chemistry. 1999;**581**: 170-175

[13] Javier L-C, Guillermo ME, Eugenio C. Polymers. 2016;**8**:171-184

[14] Deepankar D, Daniel S. Chapter In book: e-EROS Encyclopedia of Reagents for Organic Synthesis

[15] Venkat Lingaiah B, Ezikiel G, Yakaiah T, Venkat Reddy G, Shanthan Rao P. Tetrahedron Letters. 2006;**47**:4315-4318

[16] Tsao-Ching C, Shuchun JY. Synthetic Communications. 2015;**45**(5):651-662

[17] Sathaiah G, Venkat Lingaiah BP, Chandra Shekhar A, Ravi Kumar A, Raju K, Shanthan Rao P. Indian Journal of Chemistry. 2015;**54B**:953-957

[18] Shojiro K, Yoshiharu D, Kumiko K, Yasuo W. Macromolecules. 2004;**37**:5860-5862

[19] Naseeb U, Imran M, Liang K, Yuan C-Z, Zeb A, Jiang N, et al. Nanoscale. 2017;**9**:13800-13807

[20] Zhimin X, Jingyun J, Ming-Guo M, Ming-Fei L, Mu T. ACS Sustainable Chemistry & Engineering. 2017;**5**(30):2623-2631

[21] Syed I, Syed R, Yang S, Liangliang L, Asfandiyar SB, Ce-Wen N. Scientific Reports. 2017;**7**:42493-42498

[22] Shouvik M, Prasun P, Arunava G, David DD, Rahul B. Chemistry: A European Journal. 2014;**20**(33): 10514-10518

[23] Grobner T. Nephrology, Dialysis, Transplantation. 2006;**21**:1104-1108

[24] Elster AD. Radiology. 1990;**176**:225-230

[25] McDonald RJ, McDonald JS, Kallmes DF, et al. Intracranial

gadolinium deposition after contrast-
enhanced MR imaging. Radiology.
2015;**275**:772-782

[26] Leniec G, Kaczmare SM,
Typek J, Kołodziej B, Grech E,
Schilf WJ. Physics: Condensed
Matter. 2006;**18**(43):9871-9878

[27] Leniec G, Kaczmarek SM, Typek J,
Koodziej B, Grech E, Schilf W. Solid
State Sciences. 2007;**9**:267-273

Chapter 4

Current Clinical Issues: Deposition of Gadolinium Chelates

Takahito Nakajima and Oyunbold Lamid-Ochir

Abstract

Clinically available gadolinium chelate-based contrast agents (GBCAs) are divided into two groups by chelate types: linear GBCAs and macrocyclic GBCAs. The characteristic features of GBCAs are introduced in this chapter. Currently, there are two clinical issues related to the administration of GBCAs: nephrogenic systemic fibrosis (NSF) and brain deposition of gadolinium. NSF occurs in patients with chronic renal failure who had magnetic resonance imaging (MRI) examinations with GBCA injections. Frequent administrations would induce NSF, and GBCA stability would be discussed in this chapter. Linear GBCAs are more likely to be deposited in brain tissues than macrocyclic GBCAs. We present the trend of GBCA deposition or retention with our published research studies with our previous researches. We have investigated the effect of GBCAs deposited in the brain for infants.

Keywords: gadolinium-based contrast agent, nephrogenic systemic fibrosis, linear chelates, macrocyclic chelates, gadolinium brain deposition

1. Introduction

1.1 Development history

Magnetic resonance imaging (MRI) is a powerful cross-sectional diagnostic imaging modality. Its technical principle developed by Bloch and Purcell was advanced for clinical application since 1973 by Lauterbur and Mansfield [1]. MRI allows a generation of noninvasive images and a determination of detailed internal morphology and function of organs and tissues, rendering it particularly useful for detection and characterization of diseased soft tissue including solid tumors. MRI has many advantages such as the absence of ionizing radiation exposure and provides three-dimensional images with high spatial resolution and contrast. The quality of MR images, including spatial resolution, signal-to-noise ratio, and contrast-to-noise ratio (CNR), has been markedly improved in the past decades. In addition, the use of contrast agents (CAs) has been playing a crucial role in improving the detection of tumor lesions, especially brain tumors, due to a rupture of blood-brain barrier by enhancing the image contrast between normal and abnormal tissues [2, 3].

1.2 Contrast mechanism

CAs in the field of MRI alter the longitudinal (T1) and transverse (T2) relaxation rates of the surrounding water protons, therefore enhancing the image contrast in tissue of interest [4]. MRI CAs generally behave as positive CAs on T1-weighted image (T1WI) or negative CAs on T2WI based on their relaxation mechanisms. Gadolinium-based contrast agents (GBCAs) are commonly used as T1 contrast agents that have the ability to decrease T1 relaxation times of protons and work as a positive image contrast on T1WI. GBCAs have been commercially introduced since 1988 and have been globally used for more than 25 years in more than 100 million patients, and over 10 million contrast-enhanced MRI scans were annually performed [5]. These agents distribute into plasma, interstitial spaces, and extracellular spaces immediately after intravenous injection. Since most GBCAs are employed as extracellular agents, dynamic study of MRI has been performed to detect hypervascular tumors, such as hepatocellular carcinoma. The extracellular distribution of GBCAs is most effective in detection and diagnosis of disrupted blood-brain barrier in the central nervous system such as multiple sclerosis and brain tumor [6, 7].

1.3 Relaxivity

The relaxation of solvent nuclei around paramagnetic center has been described by Solomon, Bloembergen, and others [8]. Every material has proper T1 and T2 relaxation rates (1/T1, 1/T2) of water protons, and the difference of relaxivities produces the contrasts among tissues. The use of BCAs can increase both T1 and T2 relaxation rates (1/T1, 1/T2) of water protons. The observed water proton relaxation rates contribute to the contrast of the relaxation rates (1/T1, 1/T2) without GBCAs, and the increased relaxation rates (1/T1, 1/T2) are promoted using GBCAs. The increased relaxation rates of water protons are linearly related to the concentration of GBCAs within the range of clinically relevant concentrations. The relaxivity is defined as a concentration-dependent increase in relaxation rate of water protons by GBCAs in the units of $mM^{-1}s^{-1}$ [2].

$$(1/T1,2)obs = (1/T1,2)d + r1,2 [Gd] \qquad (1)$$

Protein-binding GBCAs, Gd-BOPTA (MultiHance), Gd-EOB-DTPA (Eovist), and MS-325 (Ablaber), have increased relaxivity in plasma because of their non-covalent binding to albumin which slows down the molecular rotation [2, 8]. In particular, MS-325 has an r1 relaxivity as high as $28 \pm 1\ mM^{-1}s^{-1}$ when measured at 0.47 T and 37°C in plasma [9].

1.4 Safety

Safety of GBCAs for clinical applications is another critical issue because of the reported harmful effects of Gd^{3+} in patients. Gd^{3+} ions are highly toxic in ionic form due to interference with calcium channel and protein-binding sites. This is because the ionic radius of Gd^{3+} ions is almost equal to that of Ca^{2+} and Gd^{3+} can compete with Ca^{2+} and cause toxic side effects for the biological system. Free Gd^{3+} ions accumulate in the spleen, liver, bone, and kidney, and LD50 of free Gd^{3+} ion is $0.2\ mmol\ kg^{-1}$ in mice. To prevent the toxicity of Gd^{3+} ions, chelate ligands are employed to reduce free Gd^{3+} ions. Harmful Gd^{3+} ions may still be released from some type of chelates. The mechanism of release of free Gd^{3+} from chelated CAs has been investigated. One of the hypotheses is transmetallation with other metal ions, including Zn^{2+}, Ca^{2+}, and Cu^{2+} in the serum of human body. Another hypothesis is the protonation of the ligands at low pH. These factors would cause chelate dissociation in vivo [10]. Therefore, gadolinium chelate-based MRI CAs emerged for their good safety profiles and the stability for high thermodynamic and kinetic stability.

The GBCAs are excreted from the kidney within hours after intravenous administration [11]. GBCAs are ultimately eliminated through the renal route with half-lives of 1–2 h and excreted intact in urine (more than 95% of the injected dose in 24 h). The dose of these small molecular GBCAs in clinical use is usually 100 times lower than their LD50. GBCAs used to be used as a contrast agent of MRI even for patients with chronic kidney disease (CKD). However, in 2006, nephrogenic systemic fibrosis (NSF) was reported by Grobner. Many papers reported that CKD might be the main factor of NSF [12, 13]. These days, GBCAs are not used for patients with CDK.

2. Chelate types of gadolinium-based contrast agent

GBCAs are categorized mainly into two groups: linear and macrocyclic GBCAs. In general, macrocyclic GBCAs are more stable than linear GBCAs due to higher thermodynamic and kinetic stability (**Tables 1–4**) [14, 15]. In clinical use, gado-pentetate dimeglumine, Gd-DTPA (Magnevist); gadoterate, Gd-DOTA (Dotarem); gadoteridol, Gd-HP-DO3A (ProHance); and gadodiamide, Gd-DTPA-BMA (Omniscan) have similar r1 relaxivity in the range of 3.5–3.8 mM^{-1}s^{-1} (20 MHz and 37°C) (**Tables 1–4**).

2.1 Linear chelates

Gadolinium-DTPA: Gadopentetate dimeglumine, Gd-DTPA (Magnevist), is one of the linear-type chelating agents. Gd3+ ions are covered by the polydentate ligand like a claw (**Figure 1**). The toxicity of Gd-DTPA is more than tenfold lower than the toxicity of Gd3+ ion and DTPA as a ligand. Its safety profile is very well established with low incidence of adverse effects. The risk of adverse reactions is low when then agent is administrated intravenously even up to doses of 0.03 mol/kg.

Commercial name	Chemical name	Structure	Chelate type
Magnevist	Gadopentetate dimeglumine	Gd-DTPA	Linear
Omniscan	Gadodiamide	Gd-DTPA-BMA	Linear
Dotarem	Gadoterate meglumine	Gd-DOTA	Macrocyclic
ProHance	Gadoteridol	Gd-HP-DO3A	Macrocyclic
Gadovist	Gadobutrol	Gd-DO3A-butrol	Macrocyclic

Table 1.
Representative clinical gadolinium-based contrast agents (GBCAs).

GBCAs	LD$_{50}$ (mmol/kg)	References
Gd-DTPA	8	[16]
Gd-DTPA-BMA	25	[18]
Gd-DOTA	18	[17]
Gd-HP-DO3A	<15	[19]
Gd-DO3A-butrol	25	[18]

GBCA, Gadolinium-based contrast agent; LD$_{50}$, median lethal dose.

Table 2.
Acute intravenous toxicity in rats [14].

GBCAs	Concentration (mmol/l)	Osmolality (osmol/kgH$_2$O)	Viscosity	References
Gd-DTPA	0.5	1.96	2.9	[20]
Gd-DTPA-BMA	0.5	0.79	1.4	[18, 20]
	1.0	1.90	3.9	[19]
Gd-DOTA	0.5	1.35	2.0	[19]
	1.0	4.02	11.3	[19]
Gd-HP-DO3A	0.5	0.63	1.3	[19]
	1.0	1.91	3.9	[19]
Gd-DO3A-butrol	0.5	0.57	1.4	[18]
	1.0	1.39	3.7	[20]

GBCA: Gadolinium-based contrast agent.

Table 3.
Physiochemical properties of formulations of gadolinium complexes.

GBCAS	log K (therm)	1/T1 relaxivity (1/mmol^{-1})s^{-1}	References
Gd-DTPA	22.2	3.8	[18]
Gd-DTPA-BMA	16.9	3.9	[21]
Gd-DOTA	24.7	3.5	[21]
Gd-HP-DO3A	23.8	3.7	[21]
Gd-DO3A-butrol	21.8	3.6	[18]

Table 4.
Thermodynamic stability constants and relaxivities.

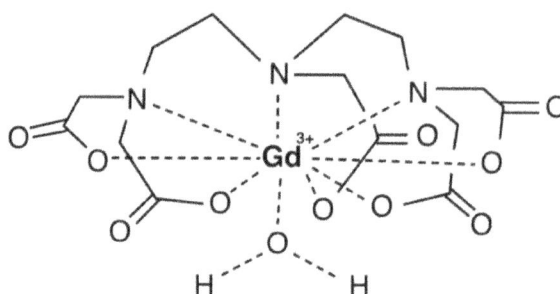

Figure 1.
Gadopentetate dimeglumine, Gd-DTPA (Magnevist).

Gadolinium-DTPA-diamides: Two extracellular contrast agents that contain neutral gadolinium chelates have entered the market: gadodiamide, Gd-DTPA-BMA (Omniscan) (**Figure 2**), and gadoversetamide (Optimark) (**Figure 3**). Both ligands are amides of DTPA and are obtained by treating the dianhydride of DTPA, corresponding to amine. These gadolinium complexes are freely soluble in water. As expected, the osmolality of the 0.5 molar solution of gadodiamide is lower (0.79 osmol/kg water) than that of the 0.5 molar solution of Gd-DTPA.

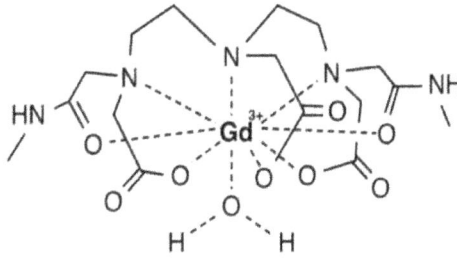

Figure 2.
Gadodiamide, Gd-DTPA-BMA (Omniscan).

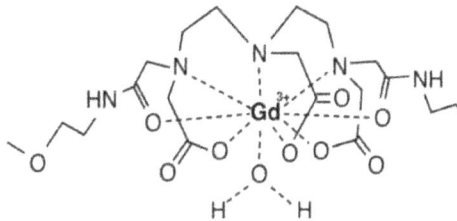

Figure 3.
Gadoversetamide (Optimark).

2.2 Macrocyclic chelates

Gadolinium-DOTA: The second generation of GBCAs contains the derivatives of the macrocyclic tetramine, 1.4.7.10-tetraazacyclododecane (cyclen). Gadoterate, Gd-DOTA (Dotarem, **Figure 4**), was the first macrocyclic gadolinium complex that was released in the market. In the macrocyclic structure, the metal-binding site within the ligand is more encapsulated, and the entropy is decreased upon metal incorporation. According to previous results, the stability of macrocyclic metal chelates is higher than that of linear complexes. The macrocyclic complexes exhibit a higher kinetic stability [10].

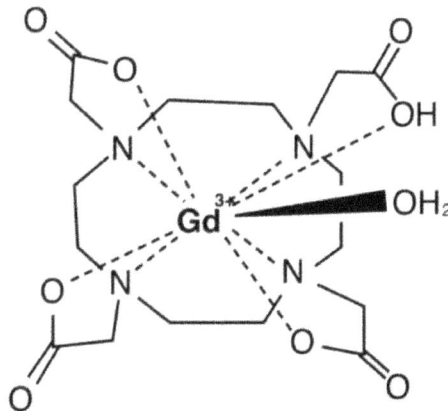

Figure 4.
Gadoterate, Gd-DOTA (Dotarem).

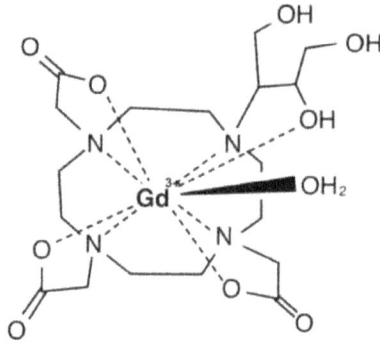

Figure 5.
Gadobutrol (Gadovist).

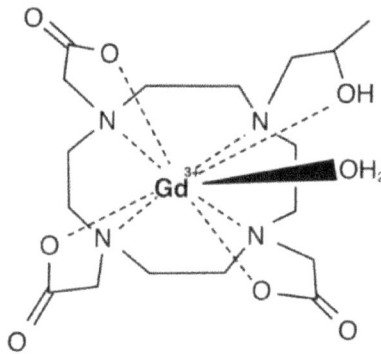

Figure 6.
Gadoteridol (ProHance).

Gadolinium-DO3A: The nonionic open-chain metal chelates and neutral macrocyclic gadolinium chelates have been synthesized. Two of them such as gadobutrol (Gadovist) (**Figure 5**) and gadoteridol (ProHance) (**Figure 6**) have been launched as extracellular MRI contrast agents. Both agents are derivatives of 1,4,7-tricarboxymethyl-1,4,7,10-tetraazacyclododecane (DO3A).

Thermodynamic stability constants and relaxivities in water at 20 MHZ and 40°C of commercially available gadolinium chelates.

3. Nephrogenic systemic fibrosis (NSF)

Nephrogenic systemic fibrosis (NSF) is a multi-systemic fibrosing disease that has been characterized by thickening and tightening of the skin and subcutaneous tissues. NSF also includes fibrosis of the skeletal muscle, lung, liver, testis, or myocardium with possible fatal outcomes.

First described in 1997, NSF was initially known as nephrogenic fibrosing dermopathy because of its classic presentation of symmetric, brawny, or erythematous indurated cutaneous plaques that develop in the setting of renal insufficiency [13, 22, 23]. Grobner's published case reports have found an association of NSF with GBCAs exposure when a meticulous chart review was performed. Currently, over 250 documents reporting NSF cases have been registered by Shawn Cowper of Yale University, which have linked 85% of its cases to gadodiamide (Omniscan) [23–25].

3.1 Pathogenesis of NSF

The exact pathogenesis of NSF is still unclear. However, there are some evidences suggesting that it likely involves the migration of CD34 and procollagen-1 positive, circulating fibrocytes from the blood to the engaged tissue as proposed by Cowper. These fibrocytes likely activate a fibrotic response through cytokine production and T-cell activation. Several reports have shown increased expression of transforming growth factor β1 and CD68-factor XIIIa within the affected skin and skeletal muscle which are also essential markers associated with wound healing and fibrosis [26, 27].

In addition, evidence of in vivo transmetallation has been provided by a preclinical trial in which rats exposed to repeated high-dose GBCAs injection developed an NSF-like skin lesion consisting of epidermal ulceration, acanthosis, dermal fibrosis, and CD34 fibrocytic infiltration with high concentrations of gadolinium in the skin. These findings were more severe with gadodiamide (Omniscan) than gadopentetate (Magnevist) [28]. The differences in conditional thermodynamic stability constants and stimulatory response of gadodiamide on fibroblasts may explain the higher incidence of NSF with gadodiamide (Omniscan). The medical literature report of NSF and this additional evidence seem to indicate a stratified risk within the class of GBCAs. The transmetallation of the gadolinium chelate would occur because of exchange of Gd3+ ions for endogenous metals (such as Zn, Cu, and Ca), and then free Gd3+ ions are released from the chelate. Patients with severe or end-stage renal disease are more likely to undergo in vivo transmetallation because of markedly prolonged clearance of GBCA. This theory has been substantiated by detection of gadolinium within tissue months after GBCA exposure. And transmetallation of Gd-DTPA with endogenous Fe(II)/Fe(III) is possible in human blood plasma. Telgmann concluded that transmetallation may be a trigger of NSF if free Fe(III) ions were accessible during a prolonged pathway of Gd complexes with linear ligands through the patient's body [29].

3.1.1 Clinical features of NSF

1. Onset: A few days to 20 years later. The early clinical symptoms of NSF include pain, pruritus, swelling, erythema (usually starts in the legs), transient alopecia, as well as gastrointestinal symptoms of nausea, vomiting, diarrhea, and abdominal pain.

2. Lately, main symptoms emerge such as nodules developing on the skin; thickened skin and subcutaneous tissues, "woody" texture and brawny plaques; joint regulation; and severe pain. Additionally, the fibrosis process involves the internal organs, e.g., muscle, diaphragm, heart, liver, and lungs. All these processes could lead to joint contractures, cachexia, and death, in a proportion of patients [12, 24].

3.2 Risk patients

High-risk patients have a history of CKD 4 and 5 (glomerular filtration rate (GFR) < 30 ml/min), dialysis, and with reduced renal function who have had or are awaiting liver transplantation. The lower risk includes patients with CKD 3 (GFR 30–59 ml/min) and children under 1 year, because of their immature renal function [22]. Younger children and elderly persons are not affected by NSF because their immune system is immature [23]. Although the pathogenesis is not revealed, the immune system might have a key role in inducing NSF.

No cases of NSF have been reported in patients with GFR greater than 60 ml/min. The role of various possible cofactors in the pathogenesis of NSF is not proven.

Pregnant patients. In the absence of specific information, it seems wise to manage pregnant patients, regardless of their renal function, in the same way as children aged under 1 year to protect the fetus.

In the use of GBCAs, serum creatinine should be measured before gadolinium contrast media administration for all patients.

3.2.1 Key points of risk patients

1. Approximately 40–50% of MRI patients receive Gd-CM.

2. The percentage of patients with CKD 3, 4, and 5 varies in different institutions.

3. Serum creatinine and estimated GFR (eGFR) are not always very accurate indicators of true GFR. In particular, acute renal failure may not be indicated by a single eGFR value.

4. Measurement of serum creatinine/eGFR before Gd-CM is mandatory before Gd-CM which have been associated with subsequent development of NSF.

5. Measurement of serum creatinine/eGFR is not necessary in all patients receiving Gd-CM.

3.3 Use of gadolinium-based contrast agents

Three GBCAs (gadodiamide, gadoversetamide, and gadopentetic acid) are currently FDA-contraindicated in patients with GFR <30 mL/min/1.73 m^2. The American College of Radiology (ACR) recommends screening GFR of any patients with known or suspected renal impairment and advises against the use of any GBCA in patients with GFR <30 mL/min/1.73 m^2, suffering acute kidney injury, or requiring dialysis [30].

3.3.1 General points

1. The risk of inducing NSF must always be weighed against the risk of denying patients gadolinium-enhanced scans which are essential for patient management.

2. In patients with impaired renal function, liver transplant patients, and neonates, the benefits and risks of gadolinium enhancement should be considered particularly carefully.

3. In patients with CKD 4 and 5 (<30 ml/min):

   • Always use the smallest possible amount of the contrast agent to achieve an adequate diagnostic examination.

   • Never use more than 0.3 mmol/kg of any Gd-CM.

   • Never use gadolinium as a contrast agent for radiography, computed tomography, or angiography as a method of avoiding nephropathy associated with iodinated contrast media.

3.4 How to choose gadolinium-based contrast agents?

There are differences in the incidence of NSF with different GBCAs, which appear to be related to differences in physicochemical properties and stability. Since macrocyclic GBCAs are preorganized rigid rings of almost optimal size to cage the gadolinium ions which have high stability compared with linear GBCAs, macrocyclic GBCAs are recommended to inject to patients with mild to moderate CKD. According to the current knowledge about the properties of the different agents and the incidence of NSF, macrocyclic GBCAs should be used for high-risk patients [31].

4. Gadolinium deposition in human body

4.1 Retention of gadolinium

The retention of gadolinium in the human body has become an issue of considerable global interest these days. Gadolinium retention was observed in the bone and in the brain in patients without renal failure [32]. The pharmacokinetics of different gadolinium chelates have been studied in healthy patients and in those with varying degrees of renal impairment. The main pathway of elimination is glomerular filtration [33]. The mean elimination half-life of GBCAs is 1.3–1.5 h. In patients with severe renal insufficiency, the half-life increases to 34.3 h [7]. Currently, it is known that gadolinium is retained in body tissues, regardless of levels of renal function or even GBCA stability [22]. Higher concentrations appear to occur in patients with renal impairment or after exposure to less stable GBCAs [34].

Our recent study found that a long-term Gd retention for GBCAs was almost unaffected by renal function [35, 36]. This finding suggested that the chemical structures of retained Gd may not be homogeneous and some Gd could be slowly eliminated after being initially retained in the tissues. Moreover, Gd retention was greater when linear GBCA was administered, than macrocyclic GBCA. However, the presence of the blood–brain barrier (BBB) likely plays a role in the mechanism of Gd retention in the brain. The mechanism of retention and the shapes of GBCAs have not been adequately revealed yet. Although injection doses should be minimized for all patients, some reports suggest that injection times would be more important than injection doses [27, 37–39].

Gadolinium has some isotopes such as 154Gd, 155Gd, 156Gd, 157Gd, 158Gd, and 160Gd as stable isotopes and 152Gd as a radioisotope. Biodistribution study of various GBCAs was performed in 1995 using a radioisotope of 153Gd. 153Gd was labeled to gadopentetate (Magnevist), gadoteridol (ProHance), gadoterate (Dotarem), and gadodiamide (Omniscan) [40]. All GBCAs were excreted from animal body within 60 min, and GBCAs in blood pool were completely disappeared. The liver, kidney, femur bone, and gastrointestinal tract still retain GBCAs until 14 days. These days the measurements of gadolinium are performed by inductively coupled plasma mass spectroscopy (ICP-MS). In our researches, either 158Gd or 160Gd is measured by ICP-MS. The technology of imaging mass would contribute to reveal the distribution of gadolinium in vivo [41, 42].

Gregory WW reported the gadolinium concentration remaining in human bone tissue after administration of 0.1 mmol/kg of two types of GBCA, Omniscan (Gd-DTPA-BMA) or ProHance (Gd-HP-DO3A), to patients undergoing hip replacement surgery. Tissue retention in bone for Omniscan (Gd-DTPA-BMA) was significantly higher than those for ProHance (Gd-HP-DO3A) measured by ICP-MS. Omniscan (Gd-DTPA-BMA) left approximately four times more gadolinium behind in bone than did ProHance (Gd-HP-DO3A). These results would indicate two important issues: (1) chelate stability and (2) affinity of GBCA to the bone. Linear GBCAs like Omniscan (Gd-DTPA-BMA) would release more Gd^{3+} ions than macrocyclic GBCAs like ProHance (Gd-HP-DO3A). Since the bone is a known natural repository for unchelated Gd^{3+} ions, free Gd^{3+} ions released from Omniscan (Gd-DTPA-BMA) would retain bone. Another reason to deposit gadolinium to bone would be the affinity of GBCAs to bone. Hydroxyapatite structure is similar to the structure of DTPA. It might be one of causes for linear GBCAs, especially Gd-DTPA or Gd-DTPA-BMA to retain bone.

4.2 Gadolinium deposition in the brain (Kanda reports)

In 2014, Kanda et al. reported unusual brain MRI findings in patients with a history of various GBCAs administration [43] (**Figures 7** and **8**). High signal intensity of the dentate nucleus and globus pallidus on unenhanced T1-weighted MR images was seen in patients with frequent administrations of GBCAs. Signal intensities on T1WI showed a positive correlation with the number of previous GBCA administrations, even in patients with normal renal function. Furthermore, the positive correlation with numbers of GBCA administrations and signal intensities on T1WI was noted only for patients with linear GBCA administrations, while no correlations were seen in patients with macrocyclic GBCA administrations [15, 44, 45]. Several researchers also reported a presence of hyperintensity in the dentate nucleus that corresponded with the number of past linear GBCAs administrations [46, 47]. Similar findings were reported in pediatric patients [30, 43].

The cause of high signals in the globus pallidus continues to be a hot topic in this field. Cadaver studies with quantitative analyses using mass spectroscopy revealed high signal intensities of the globus pallidus in patients with frequent GBCA administrations would be based on gadolinium. Although high signal intensities in the globus pallidus have been also observed in patients who have a history of liver failure, Wilson disease, Osler-Weber-Rendu disease, manganese toxicity, calcification, hemodialysis, total parenteral nutrition, and neurofibromatosis type 1, various kinds of metal ions would be deposited in the globus pallidus. However, the forms of gadolinium deposition have not been known whether chelate type,

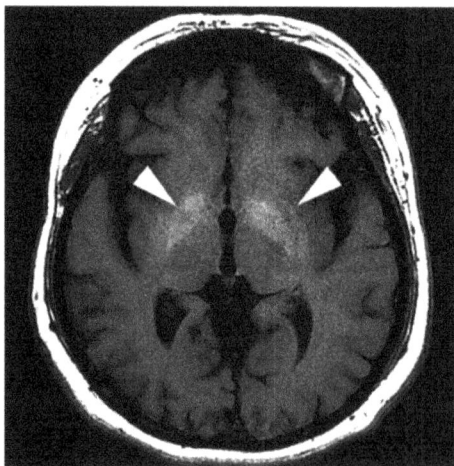

Figure 7.
Unenhanced T1-weighted MR image. Signal increase on unenhanced T1-weighted MR image in the basal ganglia, which is indicated by arrowheads.

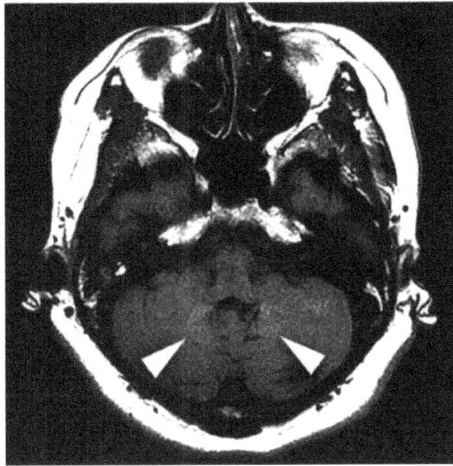

Figure 8.
Unenhanced T1-weighted MR image. Signal increase on unenhanced T1-weighted MR image in the dentate nucleus of the cerebellum, which is indicated by arrowheads.

gadolinium ions, and phosphate or other complexes. In serum or blood, gadolinium ion released from GBCAs tends to bind to phosphate quickly and to make a form of phosphate complex with gadolinium. Since phosphate complex has no signals on MRI, phosphate complex would not be a cause of high signals in the globus pallidus. In our speculation, macromolecules of gadolinium bound to protein would be one of the candidates for deposition in the brain. The reason for our speculation is that gadolinium ions bound to macromolecules show much brighter signals on T1WI than those of small molecules, which is based on the model known as the Solomon-Bloembergen-Morgan formula. Since the macromolecule rotates slowly compared with small molecules, the contact time with gadolinium and water proton can be prolonged, and the T1 shortening effect of gadolinium can be increased. The high signals in the globus pallidus would be derived from the role of macromolecules bound to gadolinium.

4.3 Effect in the central nervous system (CNS)

Gd^{3+} affects the development of infant CNS: In our previous study, we reported that Gd was transferred to pups and was retained in their brain during postnatal period. The perinatal exposure to GBCAs induced behavioral changes in mice; gadodiamide (Omniscan) had a more severe effect than gadoterate meglumine (Dotarem). All pups were separated from mothers for weaning on P21, and the total Gd in the brain of mothers and pups were measured on P28. Higher dose of Gd retention was found in mothers and pups in the linear-type group. Perinatal administration of GBCAs caused anxiety-like behaviors, disrupted motor coordination, impaired memory function, stimulated tactile sensitivity, and decreased muscle strength, especially in the gadodiamide-treated group [48]. In linear GBCA group, the total Gd retention in the brain of the mothers and the pups was higher than in the macrocyclic group. Both GBCAs were intravenously injected with 2 mmol/kg into the mothers from E15 to E19, which is the critical period for the development of neuronal circuits in fetus. According to the results of this study, we investigated Gd retention in various organs in both the mother and pup mice models [49]. Gd retentions in mother mice were consistently higher after gadodiamide (Omniscan) administration than gadoterate meglumine (Dotarem). Moreover, significantly higher Gd retention was observed in the organs of pups after whose mothers were administered gadodiamide (Omniscan) than gadoterate meglumine (Dotarem) (**Figure 9**). The results indicated that the linear GBCA affected not only the brain but also other maternal organs, such as the bone, spleen, and liver. Though the effects of maternal GBCA administration have not been reported in humans, our studies would warn the potential risk of using GBCAs in pregnant women.

4.4 Glymphatic system

The glymphatic system is discovered as a brain waste system to transport low molecular weight materials from the cerebrospinal fluid (CSF) to the interstitial fluid (ISF). The glymphatic system may also transport GBCAs to the brain. Ilif et al. examined GBCA deposition in the brain with rat by MRI [50]. When GBCA was injected into the rat subarachnoid space, it moved along the basilar artery into the brain parenchyma. In addition, Eide et al. evaluated patients who had GBCA administrations in the subarachnoid space with MRI [51]. Four hours after GBCA administration in the subarachnoid space, both the cortical and white matter of the brain showed high signal intensities, and the gadolinium entrance to the human brain through the glymphatic system was speculated. Naganawa et al. reported that on post contrast FLAIR image, the subarachnoid space and perivascular space showed increased signal intensities and GBCA transfer to the subarachnoid space and perivascular space on brain MRI of 27 subjects who had administrated GBCA before 4 h [52].

These results demonstrated that even in patients with normal renal function, intravenously administered GBCA can be transported through the glymphatic system and reach the brain. However, the association between the hyperintensity in the globus pallidus and dentate nucleus and the GBCA that is transported through the glymphatic system is still unclear. The glymphatic system transports all low molecular weight materials passively, and both the linear type of GBCA and macrocyclic GBCA are transported in the same way. However, the signal intensity of the dentate

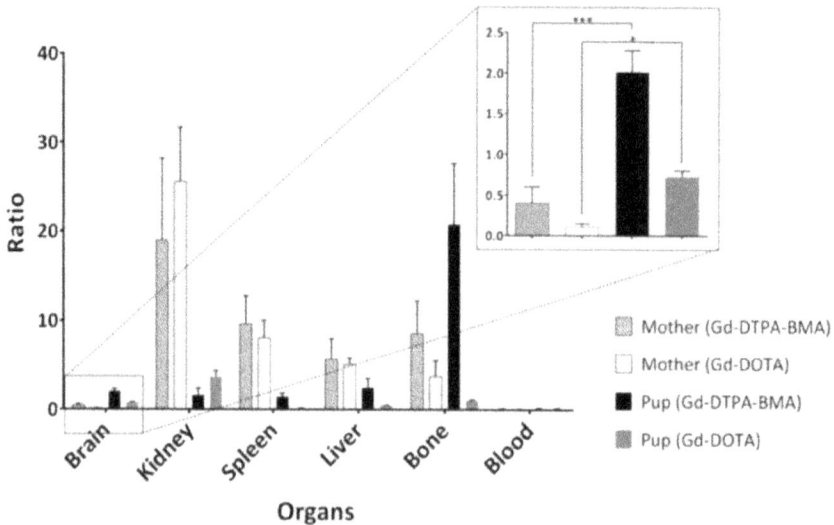

Figure 9.
Gadolinium depositions in organs of mother mice and their pups. The gadolinium concentrations in the organs of mice with the administration of linear GBCA (Gd-DTPA-BMA) were higher than those of macrocyclic GBCA (Gd-DOTA) [49].

nucleus varies according to the type of administered GBCA. In addition, the distribution of gadolinium cannot be explained by passive transportation. The accumulation of GBCA in the brain is probably due to some extent to the glymphatic system, but any association between the glymphatic system and signal hyperintensity of the dentate nucleus remains obscure.

5. Conclusion

We discussed two types of GBCAs: linear chelates and macrocyclic chelates. The macrocyclic GBCAs are more stable than the linear types because free Gd ions do not get released from the macrocyclic chelates easily in various conditions. Many preclinical and clinical studies have revealed higher deposition of Gd in the body organs in linear type than those in macrocyclic GBCAs.

Special precaution needs to be taken in cases of chronic kidney disease or patients with renal dysfunction as the only route of excretion of the GBCAs is via the kidney. Nephrogenic system fibrosis (NSF) has been noted in such renal function-impaired patients who had been administered GBCAs, especially linear types. Moreover, linear GBCAs are easy to release Gd ions from chelates. Linear GBCAs have a tendency to be deposited in the human body, including brain tissue. The use of macrocyclic GBCAs should be recommended even for patients with normal renal function.

Acknowledgements

The authors would like to thank Achmad Adhipatria P. Kartamihardja, Khongorzul Erdene, Miski Aghnia Khairinisa, and Odgerel Zorigt for their researches. This article is based on many researchers' great works and our colleagues' experiments. We also thank Nobuko Kemmotsu for editing our manuscript with English correction. Our colleagues have graduated from Gunma University Graduate School of Medicine, and they came from Asian countries to Japan to join our program, "Asian Nuclear Medicine Graduate Program (ANMEG Program)." This program was supported by the Ministry of Education, Culture, Sports, Science and Technology (MEXT) of the Japanese government. The mission of this program is to train specialists in nuclear medicine and radiology in medical fields who will become leading clinicians and researchers, both in their home countries and at an international level. The ANMEG program and MEXT supported them during their stay in Japan.

Conflict of interest

The authors have no conflicts of interest directly relevant to the content of this article.

Author details

Takahito Nakajima[1*] and Oyunbold Lamid-Ochir[1,2]

1 Gunma Graduate School of Medicine, Maebashi, Japan

2 National First Central Hospital of Mongolia, Ulaanbaatar, Mongolia

*Address all correspondence to: sojin@gunma-u.ac.jp

IntechOpen

References

[1] Haacke EM, Brown RW, Thompson MR, Venkatesan R. Magnetic Resonance Imaging: Physical Principles and Sequence Design. Vol. 82. New York: Wiley-Liss; 1999

[2] Zhou Z, Lu Z-R. Gadolinium-based contrast agents for magnetic resonance cancer imaging. Wiley Interdisciplinary Reviews Nanomedicine Nanobio-technology. 2013;**5**:1-18. DOI: 10.1002/wnan.1198

[3] Morris SA, Slesnick TC. Magnetic resonance imaging. Visual Guide to Neonatal Cardiology. 2018:104-108

[4] Runge VM. Chapter 14. Contrast media BT-clinical MRI. Clinical MRI. 2002:454-472

[5] Kanda T, Nakai Y, Oba H, Toyoda K, Kitajima K, Furui S. Gadolinium deposition in the brain. Magnetic Resonance Imaging. 2016;**34**:1346-1350. DOI: 10.1016/j.mri.2016.08.024

[6] Telgmann L, Sperling M, Karst U. Determination of gadolinium-based MRI contrast agents in biological and environmental samples: A review. Analytica Chimica Acta. 2013;**764**:1-16. DOI: 10.1016/j.aca.2012.12.007

[7] Rai AT, Hogg JP. Persistence of Gadolinium in CSF: A diagnostic pitfall in patients with end-stage renal disease. American Journal of Neuroradiology. 2001;**22**:1357-1361

[8] Nordenskiold L, Laaksonen A, Kowalewski J. Applicability of the Solomon-Bloembergen equation to the study of paramagnetic transition metal-water complexes. An ab initio SCF-MO study. Journal of the American Chemical Society. 1982;**104**:379-382. DOI: 10.1021/ja00366a002

[9] Frenzel T, Lengsfeld P, Schirmer H, Hütter J, Weinmann HJ. Stability of gadolinium-based magnetic resonance imaging contrast agents in human serum at 37 degrees C. Investigative Radiology. 2008;**43**:817-828. DOI: 10.1097/RLI.0b013e3181852171

[10] Caravan P, Ellison JJ, McMurry TJ, Lauffer RB. Gadolinium(III) chelates as MRI contrast agents: Structure, dynamics, and applications. Chemical Reviews. 1999;**99**:2293-2352. DOI: 10.1021/cr980440x

[11] Bellin M. MR contrast agents, the old and the new. European Journal of Radiology. 2006;(3):314-323

[12] Grobner T. Erratum: Gadolinium—A specific trigger for the development of nephrogenic fibrosing dermopathy and nephrogenic systemic fibrosis? (Nephrology, Dialysis, Transplantation (2006;**21**:1104-1108)). Nephrology, Dialysis, Transplantation. 2006;**21**:1745. DOI: 10.1093/ndt/gfl294

[13] Thomsen HS, Marckmann P, Logager VB. Nephrogenic systemic fibrosis (NSF): A late adverse reaction to some of the gadolinium based contrast agents. Cancer Imaging. 2007;**7**:130-137. DOI: 10.1102/1470-7330.2007.0019

[14] Sethi R, MacKeyev Y, Wilson LJ. The Gadonanotubes revisited: A new frontier in MRI contrast agent design. Inorganica Chimica Acta. 2012;**393**:165-172. DOI: 10.1016/j.ica.2012.07.004

[15] Runge VM. Macrocyclic versus linear gadolinium chelates. Investigative Radiology. 2015;**50**:811. DOI: 10.1097/RLI.0000000000000229

[16] Tóth É, Helm L, Merbach AE. In: Krause W, editor. Relaxivity of MRI Contrast Agents BT—Contrast Agents I: Magnetic Resonance Imaging. Berlin, Heidelberg: Springer Berlin Heidelberg; 2002. pp. 61-101. DOI: 10.1007/3-540-45733-X_3

[17] Doucet D, Meyer D, Bonnemain B, Doyon D, Caille JM. Gd-DOTA. Enhanced Magnetic Resonance Imaging. St. Louis, MO: Mosby; 1989. pp. 87-104

[18] Vogler H, Platzek J, Schuhmann-Giampieri G, Frenzel T, Weinmann H-J, Radüchel B, et al. Pre-clinical evaluation of gadobutrol: A new, neutral, extracellular contrast agent for magnetic resonance imaging. European Journal of Radiology. 1995;**21**:1-10. DOI: 10.1016/0720-048X(95)00679-K

[19] Tweedle MF. Physicochemical properties of gadoteridol and other magnetic resonance contrast agents. Investigative Radiology. 1992;**27**:S7

[20] Tweedle MF, Hagan JJ, Kumar K, Mantha S, Chang CA. Reaction of gadolinium chelates with endogenously available ions. Magnetic Resonance Imaging. 1991;**9**:409-415

[21] Lauffer RB. Paramagnetic metal complexes as water proton relaxation agents for NMR imaging: Theory and design. Chemical Reviews. 1987;**87**:901-927

[22] Othersen JB, Maize JC, Woolson RF, Budisavljevic MN. Nephrogenic systemic fibrosis after exposure to gadolinium in patients with renal failure. Nephrology, Dialysis, Transplantation. 2007;**22**:3179-3185. DOI: 10.1093/ndt/gfm584

[23] Deo A, Fogel M, Cowper SE. Nephrogenic systemic fibrosis: A population study examining the relationship of disease development to gadolinium exposure. Clinical Journal of the American Society of Nephrology. 2007;**2**:264-267. DOI: 10.2215/CJN.03921106

[24] Broome DR. Nephrogenic systemic fibrosis associated with gadolinium based contrast agents: A summary of the medical literature reporting. European Journal of Radiology. 2008;**66**:230-234. DOI: 10.1016/j.ejrad.2008.02.011

[25] Kanda T, Fukusato T, Matsuda M, Toyoda K, Oba H, Kotoku J, et al. Gadolinium-based contrast agent accumulates in the brain even in subjects without severe renal dysfunction: Evaluation of autopsy brain specimens with inductively coupled plasma mass spectroscopy. Radiology. 2015;**276**:228-232. DOI: 10.1148/radiol.2015142690

[26] Jiménez SA, Artlett CM, Sandorfi N, Derk C, Latinis K, Sawaya H, et al. Dialysis-associated systemic fibrosis (nephrogenic fibrosing dermopathy): Study of inflammatory cells and transforming growth factor β1 expression in affected skin. Arthritis and Rheumatism. 2004;**50**:2660-2666. DOI: 10.1002/art.20362

[27] Lever E, Sheer D. The role of nuclear organization in cancer. The Journal of Pathology. 2010;**220**:114-125. DOI: 10.1002/path

[28] Sieber MA, Pietsch H, Walter J, Haider W, Frenzel T, Weinmann HJ. A preclinical study to investigate the development of nephrogenic systemic fibrosis: A possible role for gadolinium-based contrast media. Investigative Radiology. 2008;**43**:65-75

[29] Telgmann L, Wehe CA, Künnemeyer J, Bülter AC, Sperling M, Karst U. Speciation of Gd-based MRI contrast agents and potential products of transmetalation with iron ions or parenteral iron supplements. Analytical and Bioanalytical Chemistry. 2012;**404**:2133-2141. DOI: 10.1007/s00216-012-6404-x

[30] Gale EM, Caravan P, Rao AG, McDonald RJ, Winfeld M, Fleck RJ, et al. Gadolinium-based contrast agents in pediatric magnetic resonance imaging. Pediatric Radiology. 2017;**47**:507-521. DOI: 10.1007/s00247-017-3806-0

[31] Thomsen HS. ESUR guideline: Gadolinium-based contrast media and

nephrogenic systemic fibrosis. European Radiology. 2007;**17**:2692-2696. DOI: 10.1007/s00330-007-0744-5

[32] High WA, Ayers RA, Chandler J, Zito G, Cowper SE. Gadolinium is detectable within the tissue of patients with nephrogenic systemic fibrosis. Journal of the American Academy of Dermatology. 2007;**56**:21-26. DOI: 10.1016/j.jaad.2006.10.047

[33] Staks T, Schuhmann-Giampieri G, Frenzel T, Weinmann HJ, Lange L, Platzek J. Pharmacokinetics, dose proportionality, and tolerability of gadobutrol after single intravenous injection in healthy volunteers. Investigative Radiology. 1994;**29**:709-715

[34] Ramalho J, Ramalho M, Jay M, Burke LM, Semelka RC. Gadolinium toxicity and treatment. Magnetic Resonance Imaging. 2016;**34**:1394-1398. DOI: 10.1016/j.mri.2016.09.005

[35] Kartamihardja AAP, Nakajima T, Kameo S, Koyama H, Tsushima Y. Impact of impaired renal function on gadolinium retention after administration of gadolinium-based contrast agents in a mouse model. Investigative Radiology. 2016;**51**:655-660. DOI: 10.1097/RLI.0000000000000295

[36] Kartamihardja AAP, Nakajima T, Kameo S, Koyama H, Tsushima Y. Distribution and clearance of retained gadolinium in the brain: Differences between linear and macrocyclic gadolinium based contrast agents in a mouse model. The British Journal of Radiology. 2016;**89**. DOI: 10.1259/bjr.20160509

[37] Roberts DR, Holden KR. Progressive increase of T1 signal intensity in the dentate nucleus and globus pallidus on unenhanced T1-weighted MR images in the pediatric brain exposed to multiple doses of gadolinium contrast. Brain & Development. 2016;**38**:331-336. DOI: 10.1016/j.braindev.2015.08.009

[38] Africa S. HHS Public Access. 2017;**4**:237-241. DOI: 10.1016/S2214-109X(16)30265-0.Cost-effectiveness

[39] Radbruch A, Weberling LD, Kieslich PJ, Eidel O, Burth S, Kickingereder P, et al. Gadolinium retention in the dentate nucleus and globus pallidus is dependent on the class of contrast agent. Radiology. 2015;**275**:783-791. DOI: 10.1148/radiol.2015150337

[40] Tweedle MF, Wedeking P, Kumar K. Biodistribution of radiolabeled, formulated gadopentetate, gadoteridol, gadoterate, and gadodiamide in mice and rats. Investigative Radiology. 1995;**30**:372-380

[41] Aichler M, Huber K, Schilling F, Lohöfer F, Kosanke K, Meier R, et al. Spatially resolved quantification of gadolinium(III)-based magnetic resonance agents in tissue by MALDI imaging mass spectrometry after in vivo MRI. Angewandte Chemie—International Edition. 2015;**54**:4279-4283. DOI: 10.1002/anie.201410555

[42] White GW, Gibby WA, Tweedle MF. Comparison of Gd (DTPA-BMA) (Omniscan) versus retention in human bone tissue by inductively coupled plasma mass spectroscopy. Investigative Radiology. 2006;**41**:272-278. DOI: 10.1097/01.rli.0000186569.32408.95

[43] Kanda T, Ishii K, Kawaguchi H, Kitajima K, Takenaka D. High signal intensity in the dentate nucleus and globus pallidus on unenhanced T1-weighted MR images. Radiology. 2014;**270**:1-6

[44] Gulani V, Calamante F, Shellock FG, Kanal E, Reeder SB. Gadolinium deposition in the brain: Summary of evidence and recommendations. Lancet Neurology. 2017;**16**:564-570. DOI: 10.1016/S1474-4422(17)30158-8

[45] Radbruch A, Weberling LD, Kieslich PJ, Hepp J, Kickingereder P, Wick W, et al. High-signal intensity in the dentate nucleus and globus pallidus on unenhanced T1-weighted images: Evaluation of the macrocyclic gadolinium-based contrast agent gadobutrol. Investigative Radiology. 2015;**50**:805-810. DOI: 10.1097/RLI.0000000000000227

[46] McDonald RJ, McDonald JS, Kallmes DF, Jentoft ME, Murray DL, Thielen KR, et al. Intracranial gadolinium deposition after contrast-enhanced MR imaging. Radiology. 2015;**275**:772-782. DOI: 10.1148/radiol.15150025

[47] Errante Y, Cirimele V, Mallio CA, Di Lazzaro V, Zobel BB, Quattrocchi CC. Ovid progressive increase of T1 signal intensity of the dentate nucleus on unenhanced magnetic resonance images is associated with cumulative doses of intravenously administered gadodiamide in patients. Investigative Radiology. 2014;**49**:685-690

[48] Khairinisa MA, Takatsuru Y, Amano I, Erdene K, Nakajima T, Kameo S, et al. The effect of perinatal gadolinium-based contrast agents on adult mice behavior. Investigative Radiology. 2018;**53**:110-118. DOI: 10.1097/RLI.0000000000000417

[49] Erdene K, Nakajima T, Kameo S, Khairinisa MA, Lamid-Ochir O, Tumenjargal A, et al. Organ retention of gadolinium in mother and pup mice: Effect of pregnancy and type of gadolinium-based contrast agents. Japanese Journal of Radiology. 2017;**35**:568-573. DOI: 10.1007/s11604-017-0667-2

[50] Iliff JJ, Lee H, Yu M, Feng T, Logan J, Nedergaard M, et al. Brain-wide pathway for waste clearance captured by contrast-enhanced MRI. The Journal of Clinical Investigation. 2013;**123**:1299-1309. DOI: 10.1172/JCI67677

[51] Eide PK, Ringstad G. MRI with intrathecal MRI gadolinium contrast medium administration: A possible method to assess glymphatic function in human brain. Acta Radiologica Short Reports. 2015;**4**:205846011560963. DOI: 10.1177/2058460115609635

[52] Naganawa S, Nakane T, Kawai H, Taoka T. Gd-based contrast enhancement of the perivascular spaces in the basal ganglia. Magnetic Resonance in Medical Sciences. 2017;**16**:61-65. DOI: 10.2463/mrms.mp.2016-0039

Chapter 5

Application of the Gadolinium Isotopes Nuclei Neutron-Induced Excitation Process

Igor V. Shamanin and Mishik A. Kazaryan

Abstract

The possibility of transformation of energy of fast and epithermal neutrons to energy of coherent photon radiation at the expense of a neutron pumping of the active medium formed by nucleus with long-living isomerous states is theoretically described. The channel of the nucleus formation in isomeric state as a daughter nucleus resulting from the nuclear reaction of neutron capture by a lighter nucleus is taken into consideration for the first time. The analysis of cross sections' dependence of radiative neutron capture by the nuclei of gadolinium isotopes Gd^{155} and Gd^{156} is performed. As a result, it is stated that the speed of Gd^{156} nuclei formation exceeds the speed of their "burnup" in the neutron flux. It is provided by a unique combination of absorbing properties of two isotopes of gadolinium Gd^{155} and Gd^{156} in both thermal and resonance regions of neutron energy. Conditions required for making isotope nuclei excited by forward neutron scattering on nuclei and for storing nuclei in excited states are formulated. The possibility of excess energy accumulation in the participating medium created by the nuclei of the pair of gadolinium isotopes Gd^{155} and Gd^{156} due to formation and storage of nuclei in isomeric state at radiative neutron capture by the nuclei of the stable isotope with a smaller mass is shown. It is concluded that when the active medium created by gadolinium nuclei is pumped by neutrons with the flux density of the order of 10^{13} cm^{-2} s^{-1}, the condition of levels population inversion can be achieved in a few tens of seconds. The wave length of the radiation generated by the medium is 0.0006 nm.

Keywords: gadolinium isotopes, active medium, neutron pumping, inversion of energy levels population

1. Introduction

Active medium is considered to be some matter in which it is possible to create the nucleus energy level population inversion due to radiation capture reaction and inelastic neutron scattering by the nuclei present in the matter.

The combination of nuclear transformations occurring in the matter under the influence of the neutron flux is called nuclide kinetics. Differential and integral characteristics of nuclide kinetics determine isotopic composition of the matter which was or is in the neutron field. At present, the nuclide kinetics investigation results are applied mainly in physics and nuclear reactors engineering [1] and in particular their nuclear safety. The possibility of accumulation and uncontrolled release of excess energy in neutron-absorbing materials because of potential accumulation of excess energy in isomeric states of atomic nuclei (for example, hafnium or gadolinium) comprising some of them was paid attention to [2].

2. Theoretical evaluation

Let us consider the neutron-absorbing material in which the following processes occur under the influence of neutrons: nucleus X + neutron → nucleus Y in the excited state → nucleus Y in the isomeric (metastable) state → nucleus Z in the ground state. For example: $Gd^{155} + n → Gd^{156*} → Gd^{156m} → Gd^{156}$. Nuclei Y and Z also undergo radiation capture that is they are "shot." Isomer Gd^{156m} has the half-life period of 1.3 μs and decays emitting gamma-quantum with the energy of 2.1376 MeV. In **Figure 1**, the scheme of the process is shown.

In **Table 1**, the parameters of two Gd isotopes in ground and isomeric states are presented.

Before appearing in a metastable state, Gd^{156} nucleus is in an excited state. The typical nucleus lifetime in the excited state is ~10–14 s, which is nine orders of

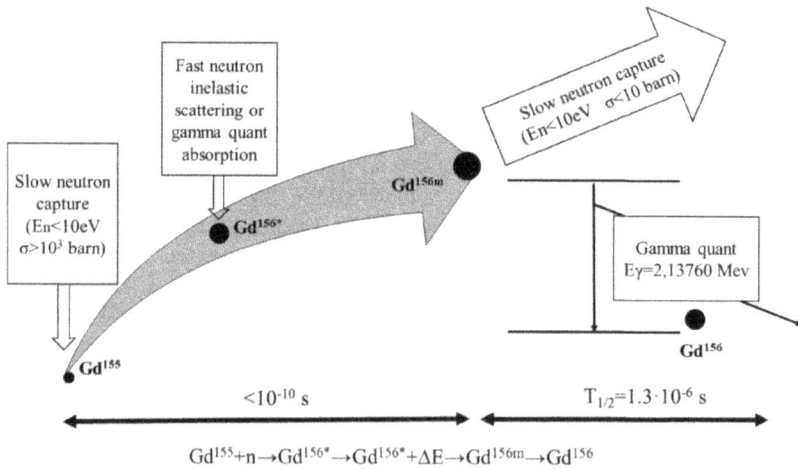

Figure 1.
Pumping of the active medium formed by gadolinium isotopes nuclei [3].

Nucleus	Half-life	Isotopic content in natural mixture	Spin and nucleus parity
Gd^{155}	Stable	14.80%	3/2-
Gd^{155m}	31.97 ms	—	11/2-
Gd^{156}	Stable	20.47%	0+
Gd^{156m}	1.3 μs	—	7-

Table 1.
Parameters of gadolinium isotopes nuclei.

magnitude longer than the nuclear interaction time. Therefore, the nucleus in the excited state can gain and conserve energy ΔE transferred to it as a result of neutron scattering on it. Energy supplied to the nucleus when the neutron dissipates on it depends on the nucleus mass—the less the nucleus mass is, the bigger energy the neutron gives to it during dissipation. In the ideal case, energy ΔE transferred at dissipation is the value equal to the difference between the energy of metastable state and energy of excitation. For the given value of the neutron energy E_n, the value of the energy ΔE transferred at inelastic scattering on the nucleus can be determined from the transcendent equation:

$$\frac{\Delta E^i}{E_n} = \frac{2A}{(A+1)^2}\left(1 + \frac{A+1}{2}\cdot\frac{\Delta E^i}{E_n} - 0.07A^{2/3}E_n\sqrt{1 - \frac{A+1}{A}\cdot\frac{\Delta E^i}{E_n}}\right), \tag{1}$$

where A is the nucleus mass number. The equation is obtained on the assumption that the transferred energy ΔE is equal to the energy of nucleus excitation from the ground state. If the nucleus has several excitation levels (i = 1, 2, 3...), the equation allows determining the value of the neutron energy E_n, providing transfer of the nucleus to the corresponding excitation level.

The number of collisions (dissipations) of the neutrons with the nuclei of active medium occurring in the medium volume unit per time unit can be determined by the following relation:

$$\Phi\sigma n_{nuc}, \tag{2}$$

where Φ is the neutron flux density, σ is the microscopic cross section of inelastic neutron scattering on the nuclei, and n_{nuc} is the number of nuclei per medium volume unit. In this case, the scattering frequency experienced by neutrons in the active medium is determined by the following relation:

$$\upsilon\sigma n_{nuc}, \tag{3}$$

where υ is the neutron velocity ($\upsilon = \sqrt{2E/m}$).

For example, to transfer nuclei of $_{54}Xe^{130}$ isotope from the ground state to the excited state, taking into account that they have three excitation levels (0.54, 1.21, and 1.95 MeV), the presence of neutrons with the energies of 0.709, 1.285, and 2.005 MeV, correspondingly, is required in the flux. To transfer nuclei of $_{10}Ne^{22}$ isotope from the ground state to the excited state, which also has three excitation levels, the presence of neutrons with the energies of 2.075, 3.747, and 4.859 MeV is required in the flux. The average neutron energy of the fission spectrum is 2 MeV. The average neutron spectrum energy of the nuclear reactor (even fast neutron reactors) is significantly lower. Besides, to transfer isotope nuclei to the excited state by direct scattering of neutrons on nuclei, it is necessary to "choose" isotopes not only with bigger specific binding energy of nucleons in the nucleus but also with small value of the neutron absorption cross section. Therefore, to accumulate nuclei in the excited state, it is reasonable to obtain them as a product of the reaction of neutron radiative capture by nuclei with a mass number smaller by one unity. The daughter nucleus is formed in the excited state and, if required, gains an additional energy due to neutron scattering on it. As a result, the daughter nucleus appears in the metastable state.

To evaluate the possibility of energy accumulation in isomeric nuclei states due to radiation neutron capture in such material, it is required to solve the differential equations system:

$$\begin{cases} \dfrac{dx}{dt} = -\sigma_1 x\Phi \\[2mm] \dfrac{dy}{dt} = \sigma_1 x\Phi - \sigma_2 x\Phi - \lambda y \\[2mm] \dfrac{dz}{dt} = -\sigma_3 z\Phi + \lambda y \end{cases} \tag{4}$$

Here $x(t)$, $y(t)$, $z(t)$ – Gd^{155}, Gd^{156m}, Gd^{156} nuclei concentration, respectively; Φ is the neutron flux density; σ is the micro-cross section of radiation neutron capture (σ_1—for Gd^{155} nuclei, σ_2—for Gd^{156m} nuclei, σ_3—for Gd^{156} nuclei), and λ is the decay constant of isomers nuclei Gd^{156m}.

Solution of the system of equations gives the formulae to determine the possibility to achieve the condition at which the nuclei concentration in isomeric state y (t) becomes bigger or equal to the concentration of nuclei in the ground state $z(t)$ influenced by neutrons with the flux density Φ to 10^{16} cm^{-2} s^{-1}:

$$\frac{y(t)}{z(t)} \approx \frac{\lambda t - (\sigma_1 - \sigma_2)\Phi t}{S\lambda}, \tag{5}$$

where

$$S = \frac{1 - (\lambda + 2\sigma_3\Phi)t}{\lambda + \sigma_3\Phi} - \frac{1 - (\sigma_1 - \sigma_2 + 2\sigma_3)\Phi t}{(\sigma_1 - \sigma_2 + \sigma_3)\Phi} + \frac{(\lambda - (\sigma_1 - \sigma_2)\Phi)(1 - \sigma_3\Phi t)}{(\sigma_1 - \sigma_2 + \sigma_3)\Phi(\lambda + \sigma_3\Phi)}. \tag{6}$$

The ratio $\frac{y(t)}{z(t)}$ is the ratio of the concentration of Gd^{156m} nuclei to the concentration of Gd^{156} nuclei. When this ratio becomes greater than 1, that means that, starting from a certain point in time, the concentration of Gd^{156m} nuclei becomes greater than the concentration of Gd^{156} nuclei. It is taken into account that the Gd^{156m} nuclei transfer to the ground state with the emission of gamma-quants that act on the Gd^{156} nuclei, transferring them to the excited metastable state.

When neutrons with the flux density $\Phi = 10^{13}$ cm^{-2} s^{-1} influence the neutrons absorber formed by gadolinium nuclei, the condition $\frac{y(t)}{z(t)} \approx 1$ is achieved within several tens of seconds. It is explained by almost unique combination of absorbing properties of two isotopes of gadolinium (Gd^{155} and Gd^{156}) in both thermal and resonant energy regions of neutrons.

3. Calculation results

An analysis of the dependences of the micro-cross sections of neutron radiative capture by gadolinium isotopes on the neutron energy, presented in [4], indicates a favorable combination of properties of two gadolinium isotopes with mass numbers of 155 and 156, consisting in preferential absorption of neutrons by the light isotope in a rather wide energy range.

The reaction cross section of the neutron radiative capture by Gd^{155} nuclei exceeds by 3–4 orders of magnitude that for Gd^{156} nuclei at neutron energies to 10 eV; the resonance integral for Gd^{155} nuclei significantly exceeds that for Gd^{156} nuclei. The production rate of Gd^{156m} nuclei is significantly higher than their "shooting" by neutrons and the rate of their ground-state transition even at flux densities of resonant and thermal neutrons of the order of $\sim 10^{13}$ cm^{-2} s^{-1}. The further increase in the neutron flux density leads to reducing the time interval after which the excess energy begins to accumulate. As a result, rapid accumulation of excess energy in the metastable state of gadolinium-156 isotope nuclei should be expected at moderate neutron flux densities.

The possibility of pumping the medium formed by hafnium nuclei with gamma-quanta was studied in [5]. External gamma-quantum flow cannot provide conditions for population inversion of metastable nuclei energy levels. To compare the ability to accumulate energy in isomeric conditions at nuclei excitation, according to the scheme presented in **Figure 1**, the stable isotope $_{72}$Hf^{178} was considered. Metastable nuclei of hafnium-178m2 form from the nuclei of hafnium-178 (stable isotope with 27.28% content in natural mixture). According to contemporary data [6–8], the energy of emitted gamma-quantum is 2.446 MeV at transition to the ground state and the half-life of 31.0 years correspond to metastable nuclei of hafnium-178m2. These parameters are different for metastable nuclei of hafnium-178m3 (higher energy level) and are equal to 2.534 MeV and 68 μs, and 1.147 MeV and 468 μs for metastable nuclei of hafnium-178m1.

The metastable nuclei of hafnium-178m2 form not only at inelastic scattering of fast neutrons on nuclei of hafnium-178, but also at radiation neutron capture by the nuclei of hafnium-177 (stable isotope with 18.6% content in natural mixture). As a result of neutron capture, the main nucleus of hafnium-178* forms in a very excited state. The excitation energy is equal to the sum of neutron-binding energy in the nucleus and neutron kinetic energy. Lifetime of the compound nucleus in a state of excitement is not more than 10^{-13} s, excitation is removed by emission of high-energy gamma-quantum, and the nucleus transfers either to the ground or one of the metastable states.

Inelastic scattering cross section on nuclei of hafnium-178 does not exceed 2.5 barns in a wide range of neutron energies, which leads to impossibility to accumulate considerable amount of energy in isomeric states only by means of inelastic scattering, even if the neutron flux density of $\Phi \sim 10^{14}$ cm^{-2} s^{-1}. The condition $\frac{y(t)}{z(t)} \geq 1$ due to only inelastic scattering will be achieved in a very big period of time. The balance of hafnium-178 nuclei in the isomeric state **m2** improves, if it is taken into account that they form as a result of radiation neutron capture by nuclei of hafnium-177. Cross section of this process is hundreds of barn for thermal neutrons and is more than 1 barn for neutrons with the energy to 100 eV. The condition $\frac{y(t)}{z(t)} \geq 1$ can be achieved in a significantly shorter period of time with account of radiation neutron capture, but if it is taken into account that as a result of neutron capture by nuclei of hafnium-178 and its isomers (the cross section of the process for thermal neutrons is tens of barn) all these nuclei disappear, the condition can be not achieved in principle.

To perform the research of excess energy accumulation in Gd, the following system to place Gd$_2$O$_3$ in the reactor core was used (see **Figure 2**).

Gd$_2$O$_3$ is placed in a cylindrical volume made of pure tungsten. Further, the cylinder is placed in the active core of the reactor unit. Uranium-graphite reactor is chosen as a reactor unit for the purpose of investigating the sample in thermal neutrons spectrum [9–13].

Several versions of the tungsten bulb with graphite reflector and without reflector are considered in the work (see **Table 2**). Graphite block serves as a reflector. Isotopic composition of Gd consists of 50% Gd^{155} and 50% Gd^{156}.

The neutronic calculation was performed using a WIMSD-5B.12 specialized program (OECD Nuclear Energy Agency). The program WIMS is applied for calculation of thermal and fast reactors. It is also successfully used for designing reactors and calculation and analysis of various effects in current reactor units. At present, the program uses the universal 69-group library of constants made on the basis of the evaluated neutron data files (ENDF, JEF, JENDL и т.д.) [11–18].

The calculation was performed for all variants of the placed sample. To compare and analyze the obtained results a preliminary calculation of the *initial spectrum* in the placement region of the sample in the reactor active core. The spectrum calculation results in the placed sample are presented in **Figures 3–7**. Changes in the

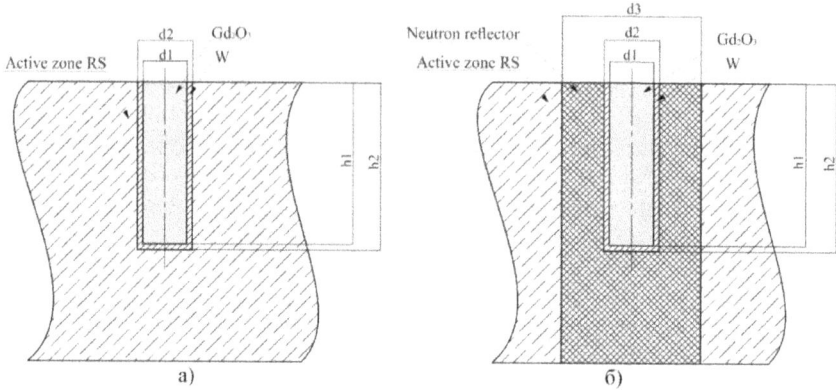

Figure 2.
The scheme of tungsten bulb placement in the reactor unit: (a) without the reflector and (b) with the reflector.

	d1, cm	d2, cm	d3, cm	h1, cm	h2, cm
Without reflector					
5_get	5	7	—	30	31
10_get	10	12	—	30	31
With the reflector					
5_get+C	5	7	30	30	31
10_get+C	10	12	30	30	31

Table 2.
Geometrical characteristics of the tungsten bulb with Gd$_2$O$_3$.

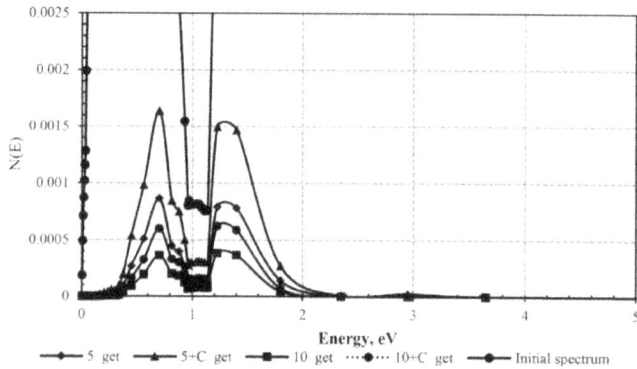

Figure 3.
The spectrum in the energy range from 0 to 5 eV.

reactor's neutron spectrum at the location of a cylinder made of Gd_2O_3 suggest that the rate of "production" of metastable Gd^{156m} nuclei will significantly exceed their "burnout" rate in the neutron field. Simultaneous fulfillment of the condition $\frac{y(t)}{z(t)} > 1$ will lead to the generation in the Gd_2O_3 volume of radiation with a wavelength of 0.0006 nm.

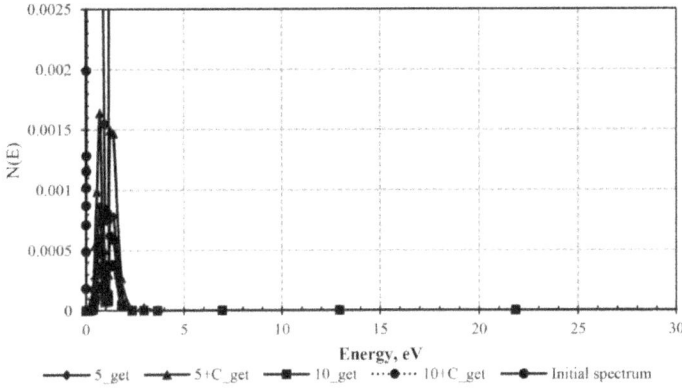

Figure 4.
The spectrum in the energy range from 0 to 30 eV.

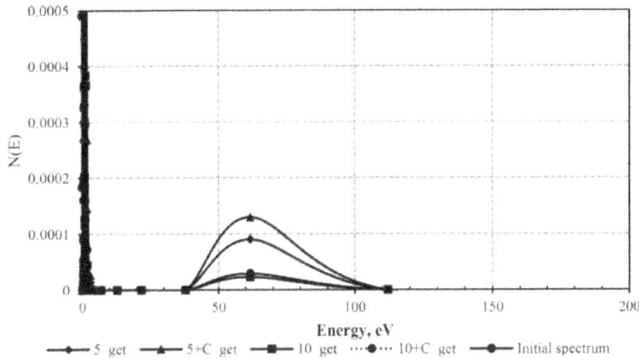

Figure 5.
The spectrum in the energy range from 0 to 200 eV.

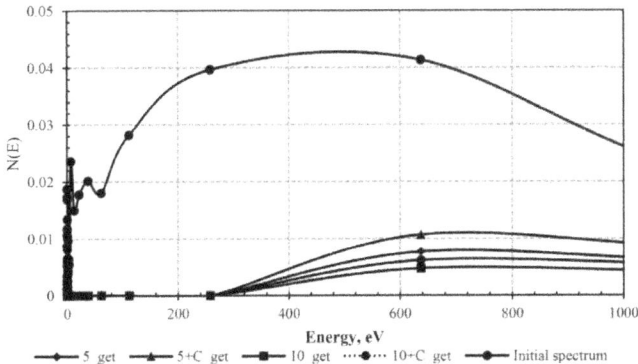

Figure 6.
The spectrum in the energy range from 0 to 1000 eV.

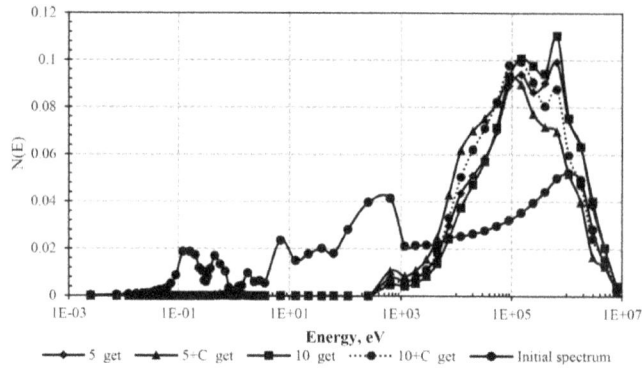

Figure 7.
Total neutron spectrum.

4. Conclusion

Accumulation of excess energy in active medium formed by the nuclei of stable isotopes of gadolinium with mass numbers of 155 and 156 due to formation of atomic nuclei in isomeric state at radiation capture of neutrons by the nuclei with the smaller mass is possible. By pumping the active medium created by gadolinium nuclei by the neutrons with the flux density Φ equal to 10^{13} cm$^{-2}\cdot$s^{-1}, the condition of the population inversion can be achieved within several tens of seconds. The wavelength of the radiation generated by the medium is 0.0006 nm. Sintered ceramics on the basis of Gd_2O_3 enriched by the 155th isotope can be considered as possible active medium. The active medium is placed in a cylindrical volume made of tungsten, which is characterized by a relatively small (to 1 barn) neutron capture cross section in a wide neutron energy range.

Author details

Igor V. Shamanin[1*] and Mishik A. Kazaryan[2]

1 National Research Tomsk Polytechnic University, Tomsk, Russia

2 P.N. Lebedev Physical Institute of the Russian Academy of Sciences, Moscow, Russia

*Address all correspondence to: shiva@tpu.ru

IntechOpen

References

[1] Shamanin IV, Bedenko SV, Pavlyuk AO, Lyzko VA. Using the ORIGEN-ARP program for calculating the isotopic composition of spent fuel of a VVER-1000 reactor. Izvestiya Tomsk Polytechnic University. 2010;**317**(4):25

[2] Shamanin IV, Kazaryan MA. Nuclide kinetics involving hafnium and gadolinium nuclei in long-lived isomeric states. Kratkie Soobsheniya po Fizike FIAN. 2017;**44**(7):48 [Bulletin of the Lebedev Physics Institute 44, 215 (2017)]

[3] Kazaryan MA et al. A mechanism for creating an inversion of populations of energy levels. Proc. SPIE 10614 (International Conference on Atomic and Molecular Pulsed Lasers XIII, Tomsk), 2018. 1061416. DOI: 10.1117/12.2303517

[4] Evaluated Nuclear Data File (ENDF). Available from: https://www-nds.iaea.org/exfor/endf.htm. Request Date is 27.04.2018

[5] Tkalia EV. Induced decay of the nuclear isomer178m2Hf and the 'isomeric bomb'. Uspekhi Fizicheskih Physics-Uspekhi. 2005;**48**(5):525

[6] Audi G, Bersillon O, Blachot J, Wapstra AH. The NUBASE evaluation of nuclear and decay properties. Nuclear Physics A. 1997;**624**:1

[7] Jain AK, Maheshwari B, Garg S, Patial M, Singh B. Atlas of Nuclear. Isomers, Nuclear Data Sheet. 2015;**128**

[8] Audi G, Konde FG, Wang M, Pfeiffer B, Sun X, Blachot J, et al. The NUBASE2012 evaluation of nuclear properties. Chinese Physics C. 2012;**36**:12

[9] Lyovina IK, Sidorenko VA. Some neutron-physical aspects of improving fuel using in water-cooled power thermal reactors VVER and RBMK. Soviet Atomic Energy. 1986;**60**:283

[10] Dollezhal NA, Ya I. Emel'anov, Channel Nuclear Power Reactor. Moscow: Atomizdat; 1980 [in Russian] (2005)

[11] Kulikov EV. The state and development prospects of NPPs with RBMK. Soviet Atomic Energy. 1984;**56**:359

[12] Romanenko VS, Krayushkin AV. Computational studies of the physical characteristics of RBMK in the transition period. Soviet Atomic Energy. 1982;**53**:367

[13] Newton TD. The Development of Modern Design and Reference Core Neutronics Methods for PBMR. Serco Assurance. Dorchester, UK: Winfrith Technology Center; 2004. Available from: https://www.answerssoftwareservice.com/resource/pdfs/enc−pmbr−paper.pdf

[14] Lindley BA, Hosking JG, Smith PJ, et al. Current status of the reactor physics code WIMS and recent developments. Annals of Nuclear Energy. 2017;**102**:148

[15] Shamanin IV, Grachev VM, Chertkov YB, et al. Neutronic properties of high temperature gas-cooled reactors with thorium fuel. Annals of Nuclear Energy. 2018;**113**:286

[16] Poveshchenko TS, Laletin NI. Method of calculating the axial diffusion coefficient of neutrons in a cell of a nuclear reactor. Atomnaya Energiya. 2016;**120**:165

[17] Altiparmakov D, Wiersma R. The collision probability method in today's computer environment. nuclear science and engineering. 2016;**182**:395

[18] Galchenko V, Mishyn A. Comparative analysis of reactor cycle neutron characteristics using different wimsd5b nuclear data libraries. Nuclear Radiation Safety. 2015;**3**(67):8

Chapter 6

Gd$_2$O$_3$: A Luminescent Material

Raunak Kumar Tamrakar and Kanchan Upadhyay

Abstract

Luminescence behavior of the Gd$_2$O$_3$ phosphor is one of the important aspects in the technology of rare earth-based inorganic phosphor materials. The structural and optical behavior of a Gd$_2$O$_3$ nanophosphor will be discussed in detail. Structural characterization of the Gd$_2$O$_3$ was carried out via X-ray diffraction and electron microscopy methods. To detail the photoluminescence behavior, the excitation and emission spectra were recorded and discussed. Thermoluminescence (TL) study and kinetic analysis of the UV- and gamma-irradiated phosphor were also carried out to determine the use of the phosphor for the dosimetric application. Tunned glow peaks were deconvoluted by applying glow curve deconvolution function, and all the trapping parameters were determined.

Keywords: Gd$_2$O$_3$, combustion synthesis method, XRD, SEM, TEM, thermoluminescence, photoluminescence

1. Introduction

These days, rare earth oxide luminescent materials have pulled in incredible consideration because of their size, shape, and phase-dependent luminescent behavior, which make them reasonable for various applications. Among various groups of crystalline materials, oxide crystal is of incredible enthusiasm because of their exceptional optical properties, for example, long fluorescent lifetime, extensive Stokes shift, positive physical and chemical properties, and also great photo-chemical stability. A few of the rare earth components and their relating oxides are of exceedingly specialized significance and are utilized in basic parts. Rare earth oxides are this sort of cutting edge materials, which are generally utilized as elite luminescent gadgets, magnets, catalyst, and other useful materials, for example, electronic, attractive, atomic, optical, and synergist gadgets [1].

Lanthanide hydroxides and oxides have effectively been examined as a result of their extensive variety of utilizations including dielectric materials for multilayered capacitors, luminescent lights and shows, strong laser gadgets, optoelectronic information stockpiles, and waveguides. Lanthanide-doped oxide nanoparticles are of unique interests as potential materials for a vital new class of nanophosphors. At the point when connected for a fluorescent naming, they present a few focal points, for example, sharp emanation spectra, long-life times, and obstruction against photobleaching in examination with ordinary natural fluorophores and quantum spots [1, 2].

Gadolinium oxide (Gd$_2$O$_3$) is one of the good choices to researchers for lumines-cence behavior because it has high refractive index (2.3), high optical transparency, great thermal and chemical stability, high dielectric consistent, and low phonon energy among the group of oxide [3–6]. Due to these positive properties, it displays various applications, for example, oxygen gas sensors, anode materials for sensors, optoelectronic gadgets, high definition TVs, medical imaging, high temperature superconducting materials, phenomenal UV light safeguard, photograph impetus, remedial impacts on malignant growth treatment-improving the impact of radiation on destructive cells while diminishing harm to typical cells, luminescent inks, paint and color sunscreen beautifiers, and luminescent materials [7].

Gadolinium oxide-based nanophosphors are observed to guarantee hopefuls in the field of superior luminescent gadgets, catalysis, and other practical gadgets dependent on their great electronic, optical, and physico-concoction reactions emerging from 4f electrons. Of course, every one of these properties could be to a great extent affected by their synthetic synthesis, precious stone structure, shape, and dimensionality. In this way, high surface region nanomaterial, which has a bigger part of deformity locales per unit zone, ought to be of enthusiasm as adsorbents in ecological remediation forms. Cost of amalgamation, effortlessness, and morphological attributes of arranged phosphor are vital parameters for their utilization in the business applications as it is basic that a self-spreading ignition course offers the best decision for the blend of Gd_2O_3 powder [2, 7].

Nanoparticles arranged by combustion synthesis have size of ~10 nm; such methodologies include the utilization of organic fuels such as urea, glycine, and so forth to start deterioration response of precursor metal salt at high temperature. The higher reactivity of littler size Gd_2O_3 particles is not simply because of the vast explicit surface region, yet in addition, because of the high concentration of low planned destinations and basic imperfections on their surface. Because of these benefits, these are sought after for different innovative applications including optoelectronic gadgets, top quality TVs, organic imaging and labeling, MRI, luminescent paints and inks for security codes, and so forth [8].

In the present work, combustion synthesis has been used for preparation of gadolinium oxide by utilizing glycerin and urea as a fuel. The union and portrayal of gadolinium oxide through various strategies have pulled in impressive consideration. The fuel and metal nitrate get deteriorated and give combustible gases, for example, NH_3, CO_2, and NO_2. At the point when the arrangement achieves a point of sudden ignition, it starts consuming and turns into a consuming strong. The ignition proceeds until the point that all the combustible substances have wore out, and it ends up being a free substance with voids and pores framed by the getting away gases amid the burning response. The entire procedure takes just a couple of minutes to yield powder of oxide. The auxiliary and optical portrayals of the incorporated powders were completed utilizing X-beam powder diffractometer. Checking electron microscopy (SEM) was utilized to show the development of crystallites, and TEM was utilized for molecular measure affirmation. Fourier Transform Infrared Spectroscopy (FTIR) range of Gd_2O_3 nanopowder was acquired by utilizing FTIR spectrophotometer (Model; MIR 8300TM) with KBr blend in the pellet shape. The Raman and X-ray photoelectron spectroscopic studies of the prepared phosphor were also carried out.

2. Materials and methods

2.1 Synthesis

Phosphor was synthesized by combustion synthesis method. Gadolinium nitrate was used as precursor solution and urea or glycine as fuel. Aqueous solution of gadolinium nitrate was prepared by dissolving suitable amount of precursor into double distilled water followed by the addition of fuel. The mixture was kept in a magnetic stirrer at 60°C and stirred for 4 h, and a transparent gel was obtained. Gel was transferred into alumina crucible and kept in a preheated furnace at 600°C. The gel mixture undergoes dehydration followed by spontaneous combustion to form Gd_2O_3 powder [1, 2]. The resulting brownish powder was heated until a controlled explosion took place yielding a very fine, white powder. Since the reaction is so rapid, the crystal growth will be highly restrained (**Figures 1** and **2**).

2.2 Material characterization

The crystallinity of the phosphor was checked by X-ray diffraction estimation. The X-ray powder diffraction information was gathered by utilizing Bruker D8 Advanced X-ray diffractometer using Cu Kα radiation. The X-beams were created utilizing a fixed cylinder, and the wave length of X-beam was 0.154 nm. The X-rays were identified utilizing a quick checking indicator dependent on silicon strip technology (Bruker Lynx Eye finder). The surface morphology of the phosphors was detected by field emission electron microscopy (FESEM) JSM-7600F. Energy dispersive X-ray examination (EDX) was utilized for compositional investigation of the phosphor. Crystal size of arranged phosphor was determined by Transmission Electron Microscopy (TEM) utilizing Philips CM-200. Raman spectra were recorded by Jobin-Yvon, France, Ramnor HG-2S Spectrometer with Ar-Laser with 4 W control having goals of 0.5 cm^{-1} and wave number exactness of 1 cm^{-1} over 5000 cm^{-1}. XPS investigation was performed in a VG instrument with a CLAM2 analyzer and a twin Mg/Al anode. The weight pressure in the investigation chamber was roughly 9×10^{-10} mbar. The estimations were done with unmono-chromated Al Kα photons (1486.6 eV). The intensity of the X-ray source was kept steady at 300 W.

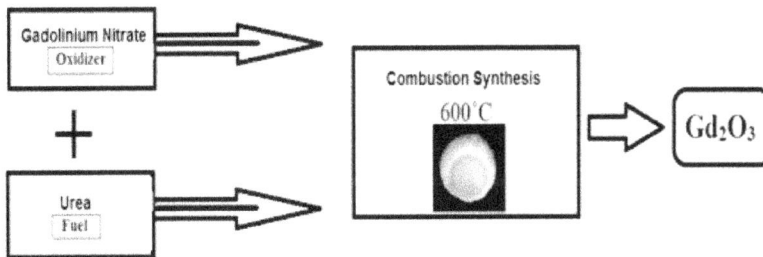

Figure 1.
Flow chart of synthesis of Gd₂O₃ with urea (reproduced from [1]).

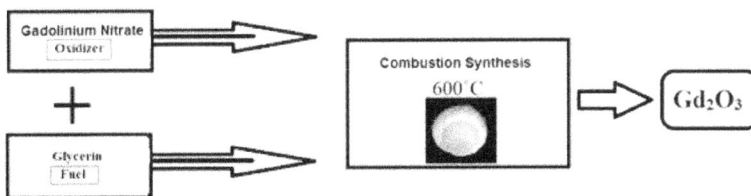

Figure 2.
Flow chart of synthesis of Gd₂O₃ with glycerin (reproduced from [2]).

3. Result and discussion

3.1 XRD result

The XRD pattern of the Gd_2O_3 sample is shown in **Figure 3**. The diffraction patterns are well matched with standard JCPDS card no. 43-1015, indicating that the sample of Gd_2O_3 phosphor is in the pure monoclinic phase. The particle size was calculated by the Scherer formula [7]

$$D = \frac{k\lambda}{\beta Cos\theta} \qquad (1)$$

where D is the volume weighted crystallite size, k is the shape factor (0.9), λ is the wavelength of Cu $K\alpha1$ radiation, β_{hkl} is the instrumental corrected integral breadth of the reflection (in radians) located at 2θ, and θ is the angle of reflection (in degrees) utilized to relate the crystallite size to the line broadening. The average crystallite size of Gd_2O_3 nanoparticles was found to be in the range of 8–10 nm for both the fuels. No impurity peaks or other possible phases of Gd_2O_3 were observed. Further, the strong and sharp diffraction peaks confirm the high crystallinity of the products.

3.2 Surface morphology

The scanning electron microscopy (SEM) was utilized as a focused ray of high energy electrons to produce an assortment of signs at the crystalline surface.

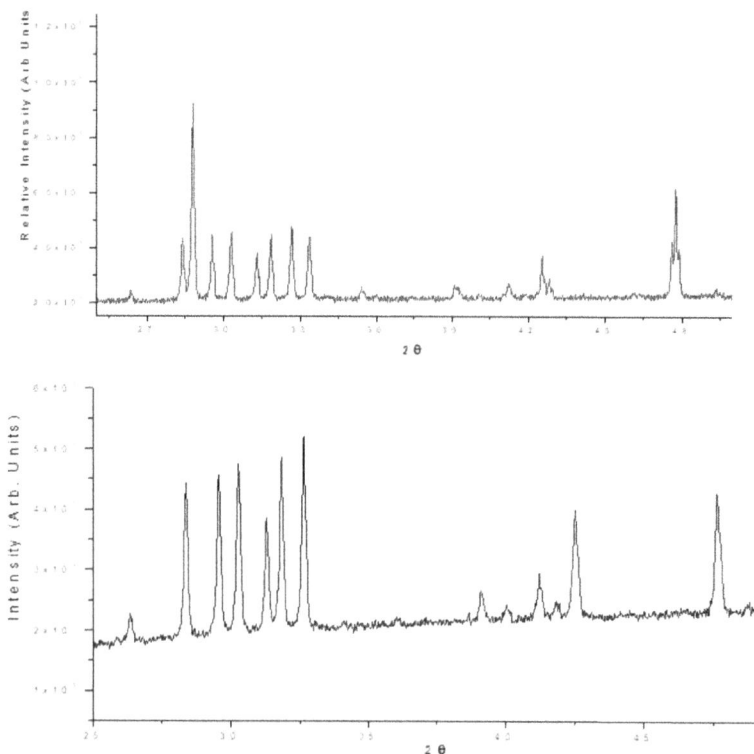

Figure 3.
XRD patterns of Gd_2O_3; (A) glycerin fuel and (B) urea fuel.

The signs that get from electron and sample interaction uncover data about the example including outer morphology, elemental composition, and crystalline structure and introduction of materials making up the example. The SEM is likewise fit for performing examinations of chose point areas on the example; this methodology is particularly valuable in subjectively or semi-quantitatively deciding synthetic structures. **Figure 4** demonstrates the SEM micrographs of the Gd_2O_3 arranged by combustion synthesis method utilizing urea and glycine as a fuel. The black and white SEM micrograph of the prepared powder indicates that all the particles are looking like agglomerated in homogeneously in different shapes/sizes of the order of nano range.

3.3 TEM result

Transmission electron microscopy (TEM) is an imaging system whereby a light emission is engaged onto an example making a broadened form show up on a fluorescent screen or layer of photographic film or to be distinguished by a CCD camera. The main commonsense transmission electron magnifying instrument was built by Albert Prebus and Lames Hillier at the college of Toronto in 1938 utilizing ideas grew before by Max Knoll and Ernst Ruska. The particle size of the system was determined by high resolution transmission electron microscopy (HRTEM). It is a phase differentiated imaging process because the image formed is due to the scattering of electron waves through a thin surface. In **Figure 5**, HRTEM micrograph demonstrates a Gd_2O_3 nanocrystal with a width of 8–10 nm seen all through the particle for both fuels [1, 2, 7].

3.4 Energy dispersive X-ray analysis (EDX)

Elemental investigation of the prepared samples is generally determined by EDX analysis. The spectrum shows the relation between the X-ray energy, which lies in between 10 and 20 eV, and the number of counts per channel by a plot between them in X and Y axes, respectively. An X-ray line is expanded by the reaction of the framework, delivering a Gaussian profile. Energy resolution is characterized as the full width of the crest at half maximum height (FWHM). In the spectrum of both the Gd_2O_3 samples, intense peak of Gd and O is present, which confirms the formation of Gd_2O_3 phosphor (**Figure 6**). For EDX analysis, the entire area of the black and white SEM micrographs was analyzed with EDX mapping and spectrum. The

Figure 4.
Scanning electron microscope image of Gd_2O_3 phosphor: (A) glycerin fuel and (B) urea fuel.

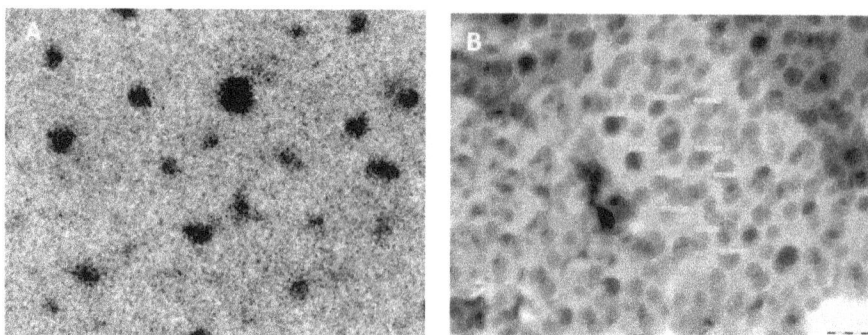

Figure 5.
Transmission electron microscope image of Gd_2O_3 phosphor: (A) glycerin fuel and (B) urea fuel.

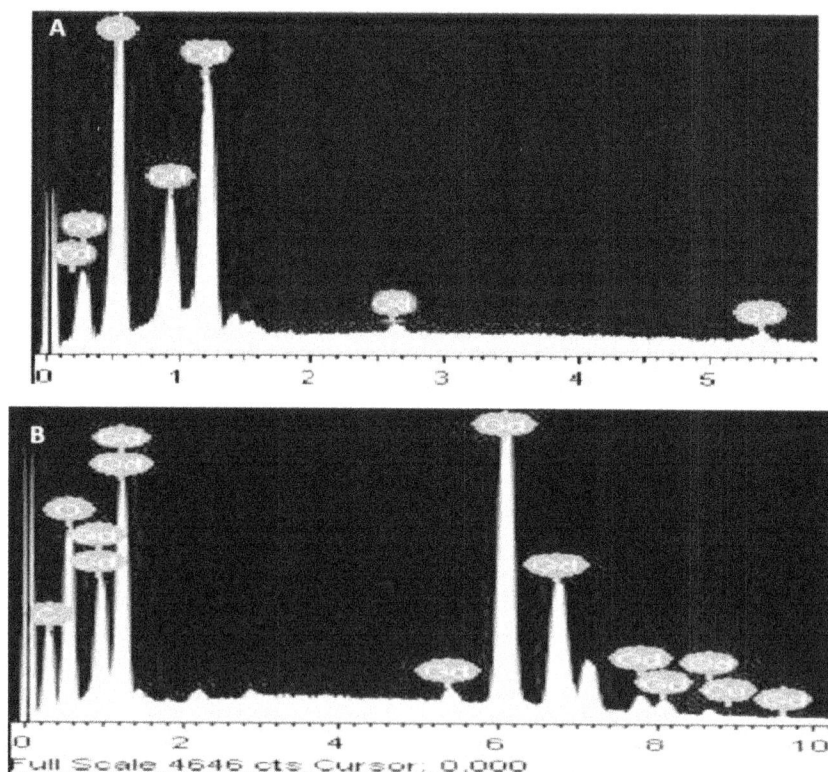

Figure 6.
EDX spectra of Gd_2O_3 phosphor: (A) glycerin fuel and (B) urea fuel.

EDX mapping measurements were carried Gd_2O_3 powders to analyze the composition of the clustered particles [1, 2].

3.5 X-ray photoelectron spectroscopy (XPS)

XPS is a surface compositional investigation system that can be utilized to examine the surface chemistry of a material in its as-formed state, or after some treatment, for instance: cracking, cutting, or scratching in air or UHV to uncover the bulk chemistry, ion beam etching to wipe off a few or the majority of the surface defilement or to purposefully uncover further layers of the sample inside and out profiling XPS, presentation to warmth to think about the progressions because of warming, introduction to receptive gases or arrangements, introduction to particle bar embed, and introduction to bright light. The synthetic organization of Gd_2O_3 nanoparticles was contemplated with X-ray Photoelectron Spectroscopy (XPS), and the test information was broken down utilizing bend fitting. The Gd (3d) level comprised of a turn circle split with the Gd (3d)5/2 top is found at 1186.74 eV (**Figure 7**). The line shape and pinnacle positions are in great concurrence with prior distributed information on Gd_2O_3 powder squeezed into the sheet [1].

3.6 Raman spectroscopy

To understand the molecular structure, Raman effect has been used, and the obtained Raman data can be compared with the infrared spectra. Raman spectroscopy is very informative to illustrate the structure of the phosphor. It is a nondestructive device to investigate vibrational, rotational, and other low recurrence modes in the frameworks under study. **Figure 8** demonstrates the Raman spectra of Gd_2O_3 obtained by combustion synthesis method. The spectra were recorded at room temperature with an excitation wavelength of 633 nm He-Cd laser. An broad and intense Raman crest at 340 cm^{-1} along with less extreme peaks was seen at 375, 395, 424, and 451 cm^{-1}. The outcomes are in great concurrence with the recently distributed Raman spectroscopic examinations on Gd_2O_3 nanoparticles [1].

3.7 Photoluminescence spectra and CIE diagrams of Gd_2O_3 phosphors

The emission spectra of Gd_2O_3 phosphor prepared with both the fuels have emission peaks at UV and visible region. A slight variation in peaks was observed in emission peaks for both phosphors. The emission spectra of Gd_2O_3 phosphor prepared by combustion synthesis method have peak at UV region in between 317 and 399 nm along with weak blue band around 450–494 nm, green around 515–586 nm, and red emission centered at 616–625 nm (**Figure 9**).

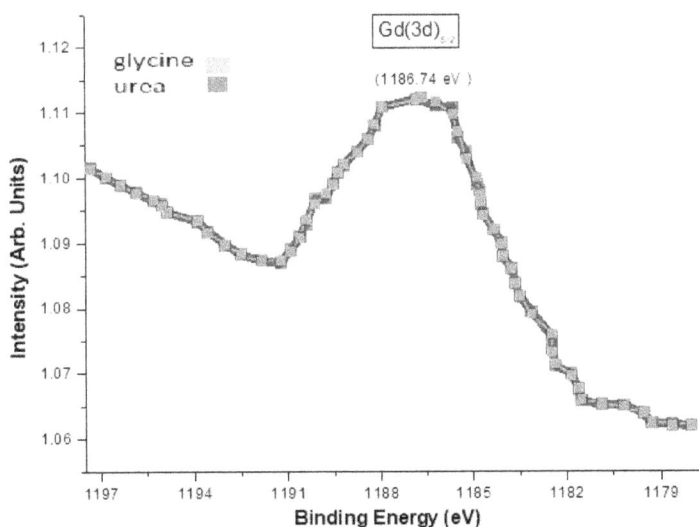

Figure 7.
The Gd (3d) XPS spectrum of Gd₂O₃ nanocrystals (reproduced from [1]).

$^6P_{7/2} \rightarrow {}^8S_{7/2}$ transition is responsible for the UV emission centered at 317 nm, whereas the visible emissions are due to transition from ^6G_J state of Gd^{3+} [8]. The presence of oxygen vacancy and interstitials also contributes in modified photoluminescence response for oxide-based system [8]. Transition from ^6G_J state of Gd^{3+} ion and $^6G_J/^6P_J$ transition is responsible for green and red emissions, respectively (**Figure 10**).

To determine the specific color produced by the prepared Gd_2O_3 phosphor, CIE coordinate diagram was prepared by using MATLAB 7.10.0 (R2010a) software. The CIE coordinates for combustion synthesized Gd_2O_3 phosphor were found $X = 0.207$ and $Y = 0.206$, which resemble with blue color. Effect of annealing on the produced color was determined by the CIE coordinates of Gd_2O_3 phosphor annealed at 900°C. It was observed that the X and Y coordinates for the annealed sample have same values as freshly prepared samples, and only the change in intensity was observed after annealing (**Figure 9**) [9].

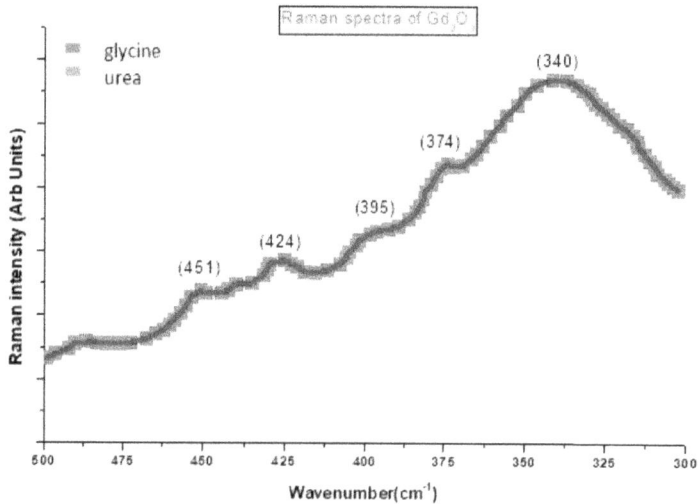

Figure 8.
Raman spectra of Gd_2O_3 nanoparticles (reproduced from [1]).

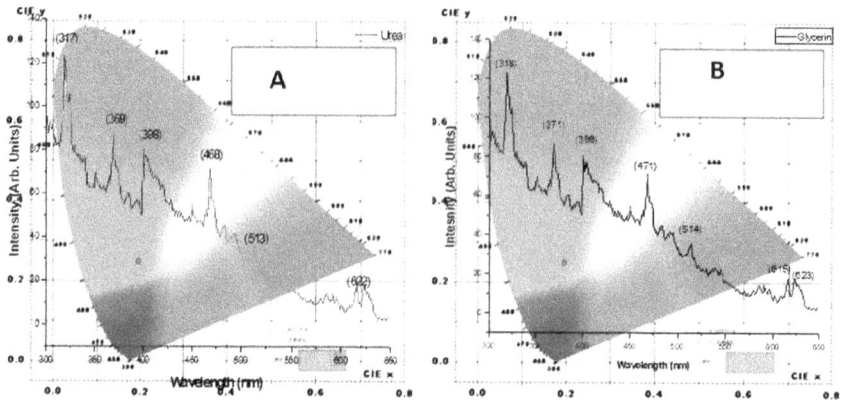

Figure 9.
Emission spectra of pure Gd_2O_3 phosphor: (A) urea and (B) glycerin (reproduced from [7]).

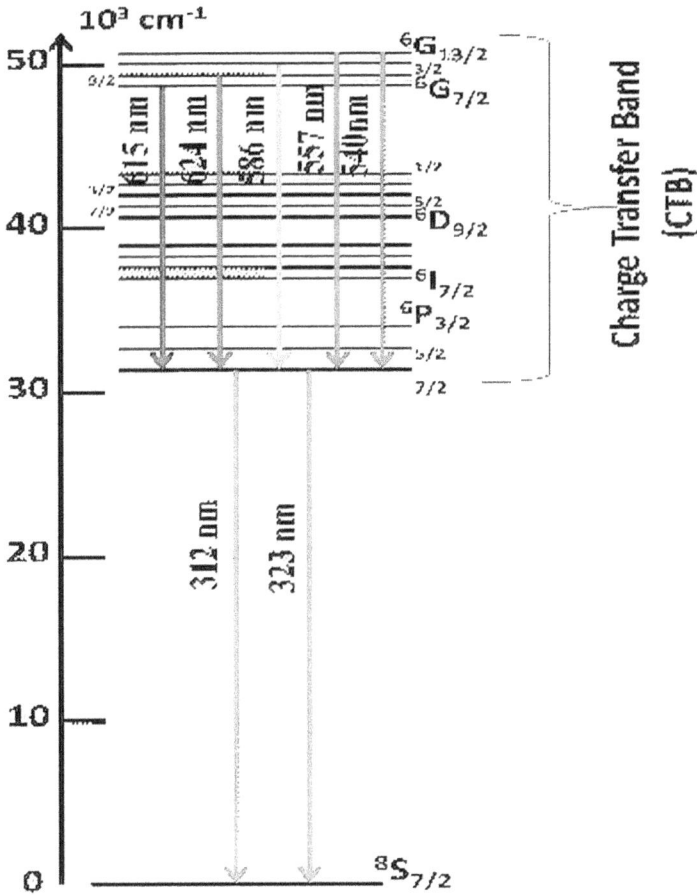

Figure 10.
Energy level diagram for emission transitions for pure Gd₂O₃ phosphor (reproduced from [7]).

3.8 Thermoluminescence study of pure Gd$_2$O$_3$ phosphor

The TL response of the Gd$_2$O$_3$ phosphor was recorded under 254 nm UV exposure and ^{60}Co gamma exposure for the phosphors prepared by both urea and glycine
fuel. The TL glow curve of phosphors prepared with both the fuels was recorded
under 254 nm UV exposure immediately after 5 min exposure time at 6 Cs^{-1} heating
rate. For the combustion synthesized phosphor, the TL glow peak was found at
103°C and 111°C for urea and glycine fuels, respectively. For 1 kGy gamma exposure
at 6 Cs^{-1} heating rate, the TL glow peak was found at 232°C and 221°C (**Figure 11**)
for urea and glycine fueled phosphors, respectively [10].

Chen's peak shape method was used to determine all the kinetic parameters
including order of kinetic, activation energy, shape factor, and so on [4–6, 11]
(**Table 1**). The activation energy for TL glow curve for combustion synthesized
both phosphors has 0.66 eV for UV exposure and 0.71 and 0.72 eV for gamma
exposure. Due to gamma exposure, deeper traps were formed, which are responsible for the higher activation energy value. The phosphor follows second-order
kinetics as the obtained shape factor value for UV exposure and gamma exposure
was in the range of 0.49–0.52 and 0.50–0.54, respectively, which is near to 0.52 for
second-order kinetics (**Table 2**).

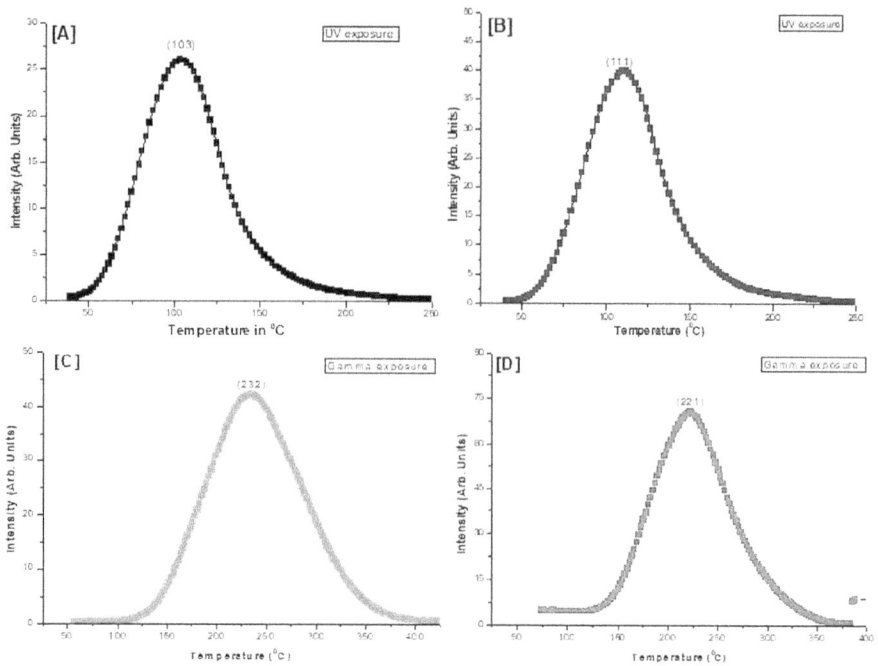

Figure 11.
TL glow curve for 6 Cs^{-1} heating rate and 10 min UV exposure with (A) urea fuel and (B) glycerin fuel. TL glow curve for 6°Cs^{-1} heating rate and 1 kGy gamma exposure (C) urea fuel and (D) glycerin fuel (Reproduced from [8]).

		Monoclinic	
		Urea	**Glycerin**
For 275 nm	X	0.207	0.207
	Y	0.206	0.206

Table 1.
X, Y coordinates for CIE diagram for Gd_2O_3 phosphor.

Exposure	Fuel	μ_g	E (eV)	S (s^{-1})
UV (254 nm)	Urea	0.49	0.66	1.4×10^{10}
	Glycerin	0.52	0.66	1.5×10^{10}
Gamma (1kGy)	Urea	0.54	0.71	9.5×10^{9}
	Glycerin	0.50	0.72	9.7×10^{9}

Table 2.
Trapping parameters for optimized TL glow curve.

4. Conclusions

The study confirms that the combustion synthesis method is suitable for large-scale production of the phosphor in minimum time. Structural characterization shows that the phosphors have monoclinic structure with particle size in the range of 8–12 nm. Phosphor synthesized by this method has homogenous particle size distribution. X-ray Photoelectron Spectroscopy (XPS) show the Gd (3d) level consists of a spin orbit split doublet, with the Gd (3d)5/2 peak is found at 1186.74 eV. Raman spectra was recorded with excitation of 633 nm wavelength, we found a broad and intense Raman peak at 340 cm^{-1} along with less intense peaks were observed at 375, 395, 424 and 451 cm^{-1}. The emission spectra have peaks in all UV and visible regions. So that, the phosphor may behave as white light emitting phosphor, which was further confirmed by its CIE coordinates. The CIE coordinates for combustion synthesized Gd_2O_3 phosphor have values $X = 0.207$ and $Y = 0.206$, and for the glycine synthesized Gd_2O_3 phosphor $X = 0.209$ and $Y = 0.207$. The values of CIE coordinates show that the Gd_2O_3 phosphor prepared by combustion synthesis emits blue color. The TL studies of both the phosphors were carried out under UV and gamma exposure. The activation energy for 0.66 eV and 0.71–0.72 eV for UV exposure and gamma exposure respectively. The value of shape factor µg for all the TL analysis was found in between 0.45 and 0.54, which shows that the phosphors follow the second-order kinetics.

Acknowledgements

We are very grateful to NIT Raipur for XRD characterization and also thankful to Dr. Mukul Gupta for his co-operation. We are thankful to SAIF, IIT, Bombay and IIT Delhi for other characterization such as SEM, TEM, FTIR, and EDX.

Conflict of interest

The authors declare there is no conflict of interest.

Author details

Raunak Kumar Tamrakar[1*] and Kanchan Upadhyay[2]

1 Department of Applied Physics, Bhilai Institute of Technology (Seth Balkrishan Memorial), Durg, Chhattisgarh, India

2 International and Inter University Centre of Nanoscience and Nanotechnology, Mahatma Gandhi University, Kottayam, Kerala, India

*Address all correspondence to: raunak.ruby@gmail.com

IntechOpen

References

[1] Tamrakar RK, Bisen DP, Sahu IP, Brahme N. Raman and XPS studies of combustion route synthesized monoclinic phase gadolinium oxide phosphors. Advance Physics Letter. 2014;**1**(1):1-5

[2] Tamrakar RK, Bisen DP, Sahu IP, Brahme N. Structural characterization of combustion synthesized Gd_2O_3 nanopowder by using glycerin as fuel. Advance Physics Letter. 2014;**1**(1):6-9

[3] Tamrakar RK, Bisen DP, Sahu IP, Upadhyay K, Sahu M. Structural characterization of Gd_2O_3 phosphor synthesized by solid-state reaction and combustion method using X-ray diffraction and transmission electron microscopic techniques. Journal of Display Technology. 2016;**12**(9):921-927

[4] Chen R. Thermally stimulated current curves with non-constant recombination lifetime. British Journal of Applied Physics. 1969;**2**:371-375

[5] Chen R, Kirsh Y. The Analysis of Thermally Stimulated Processes. Oxford, New-York: Pergamon Press; 1981

[6] Chen R, McKeever SWS. Theory of Thermoluminescence and Related Phenomena. London, NJ, Singapore: World Scientific Publications; 1997

[7] Tamrakar RK, Bisen DP, Upadhyay K, Sahu M, Sahu IP, Brahme N. Comparison of emitted color by pure Gd_2O_3 prepared by two different methods by CIE coordinates. Superlattices and Microstructures. 2015;**88**:382-388

[8] Tamrakar RK, Bisen DP, Upadhyay K, Sahu IP. Comparative study of thermoluminescence behaviour of Gd_2O_3 phosphor synthesized by solid state reaction and combustion method with different exposure. Radiation Measurements. 2016;**84**:41-54

[9] Tamrakar RK, Bisen DP, Brahme N. Comparison of photoluminescence properties of Gd_2O_3 phosphor synthesized by combustion and solid state reaction method. Journal of Radiation Research and Applied Science. 2014;**7**:550-559

[10] Tamrakar RK, Bisen DP, Brahme N. Characterization and luminescence properties of Gd_2O_3 phosphor. Research on Chemical Intermediates. 2014;**40**:1771-1779

[11] Chen R, Pagonis V. Thermally and Optically Stimulated Luminescence: a Simulation Approach. Chichester: Wiley; 2011

www.ingramcontent.com/pod-product-compliance
Lightning Source LLC
Chambersburg PA
CBHW081232190326
41458CB00016B/5755